FROM RESEARCH
TO PRINTOUT:

Creating
Effective Technical
Documents

JOHN H. WHITE, Ed.D.

TechWrite, Inc.

SME Press   New York   1997

Copyright © 1997 by   The American Society of Mechanical Engineers
                        345 East 47th Street, New York, NY 10017

Library of Congress Cataloging-in-Publication Data

From research To printout: creating effective technical documents /
John H. White
        p.   cm.
    Includes bibliographical references and index.
    ISBN 0-7918-006 0 / V
    1.   Technical writing.
    2.   Technical reports.
    I.   Title.
    T11.W448   1997   97-35091
    808'.0666——dc21   CIP

Dedicated to the loving memory of my father – James M. White, Sr.

CONTENTS

PREFACE

You have just been informed that you are to research and write a technical report on your competitor's latest manufacturing process that is providing its product at a lower cost and higher quality than your company can currently produce. The report is to be completed, duplicated, and distributed to the product development section in four days. Also, during the past three months, your company has gone through a painful downsizing process that included retirement incentives, reorganization, and elimination of all department secretarial and support positions. The downsizing has drained your department of many of the best and most experienced engineers and technicians who would normally help with research projects. The elimination of all department secretaries means that the remaining engineers and technicians must word -process or desktop publish their own technical documents. You soon realize that this will be a solitary project where you will be totally responsible for almost all phases of the project, from research to writing to word processing — welcome to the corporate world of the late 1990s.

What This Book Is

From Research to Printout: Creating Effective Technical Documents is a comprehensive publication designed to help engineers, scientists, and technicians (hereafter called technical professionals) research, develop, and produce quality technical documents. Through narratives and actual industry examples, the publication provides the reader with practical information that can be used on a daily basis in the workplace. This publication approaches the development of technical documents in a tripartite process: first, how to research the respective document topic using the latest available computer technologies and services to sift through the mountain of available data and information; second, how to outline and write the technical document and the necessary activities to enhance it through the editing process; and third, how to use today's sophisticated computers and software to produce a high-quality technical document that will get the reader's attention.

What This Book Is Not

Although this publication does cover the common grammatical errors made by technical professionals in the engineering, scientific, and technical fields, it will not dwell on grammar as taught in English 101 or business communications courses. This publication assumes the reader has had some formal writing instruction, including English at the high school and college level. Also, hundreds of excellent publications are available for this purpose.

This book will not attempt to force the reader to use a particular writing style, but will focus on those aspects of technical writing that cause technical professionals the greatest degree of problems and common errors. The publication focuses on problems that can be immediately addressed or rectified.

Why This Book Is Important

The past decade of global competition and technological change has caused many significant changes in the workplace. Perhaps the most paramount is the recent practice of downsizing the nation's and world's industries and businesses. Those employees fortunate enough to hold onto their jobs generally find that they must absorb the workload of their colleagues who lost their jobs. Also, most of today's technical professionals are required to word-process or desktop-publish their own technical documents. Additionally, because of the arrival of new computer and information technologies, the amount or volume of available data and information has increased to record levels. In turn, the technical professionals remaining in the workplace must now review, interpret, and generate quality technical reports at an unprecedented pace. Also, the virtual explosion of information means that for today's technical document consumers (e.g., supervisors, customers, colleagues, etc.), reading time is at a premium, making document layout and appearance a critical factor.

In short, this book is perhaps the first of a new generation of technical writing publications that combine the importance of research, writing, document development, and production in one publication with the goal of helping the technical professional develop a quality technical document that will be read and understood by the intended audience. Achievement of this goal will undoubtedly help the audience become professionally competitive.

Summary of Chapters

Following a logical format, the publication is divided into the following 13 chapters.

- **Chapter 1 — "Overview of Technical Writing in the Workplace"**
  This short review on the role of technical writing and professional preparation covers a brief history of technical writing and how its importance has increased during the past decade as related to the advancement of the technical professional. It also specifies the tools required to help the technical professional research, develop, and produce quality technical documents.

- **Chapter 2 — "Research"**
  This chapter includes a review of the major research sources from the traditional university research libraries to the popular on-line services such as CompuServe to the ever-growing Internet. Coupled with the review is some helpful information on which research sources are best suited for the technical professional and some helpful hints on how to maximize benefits from these sources. Also, this chapter provides information on the purposes of writing and matching the writing to the reading level of the target audience. Information is included on commercially available readability computer programs and commercial tests or assessments that help determine the grade equivalency of the target audience.

- **Chapter 3 — "Outlining and Writing the Technical Document"**
  One of the most critical stages in the technical writing process is the need to outline and actually start writing the document. This chapter provides helpful tips on how to use computer-generated outlines to help with the process, how to stay with the main thesis, how to overcome writer's block, and how to develop that all-important first draft. It also includes common writing problems of technical professionals, ranging from proper choice of words to grammar.

- **Chapter 4 — "Producing the Technical Document With a Personal Computer"**
  This chapter explains the advantages and disadvantages of producing the technical document in word-processing and desktop-publishing programs, along with the advantages of employing today's sophisticated spelling and grammar checkers, and how to get the most out of these useful tools. Helpful hints and information on how to obtain maximum use from word-processing and desktop-publishing programs in document layout and production are included. There is also a discussion on the world of computer graphics and how to use these graphics to enhance the look and support the text of the technical document. The need to regularly back up the computer files in the very likely event of a system failure is discussed.

- **Chapter 5 — "Editing the Technical Document"**
  This chapter includes all critical elements related to editing of the technical document, ranging from editing by the author and colleagues to using proofreaders' marks to clarifying the text. Tips are given on enhancing order and logic and on how today's computers can help streamline this critical stage. Additionally, tips and information are provided on the importance of ensuring that the document layout is correct and consistent, reviewing accuracy of quoted material, and reviewing corrections of tables, formulas, and other data. This also includes information on proper use of copyrighted material in a technical document to ensure that it is used legally and credit is given to the original writer.

Chapter 6 —"Electronic Documents: Facsimiles (Faxes) and Electronic Mail (E-Mail)
This chapter includes an interesting history of fax technology and expectations for the future. It also includes tips on how to get the most out of this technology, which is used in almost every business internationally. In addition to fax technology, this chapter includes information on electronic mail (e-mail) that is now the most widely used form of communication in today's workplace. It is also a form of communication that when used improperly encourages communication that is more suitable for the computer hobbyist than the professional. This

chapter focuses on how to use this important technology in an efficient manner to help project a professional image and ensure that the e-mail message is not misunderstood by the intended reader. This chapter also includes a discussion of the various types of e-mail and provides helpful hints on developing a readable and attractive e-mail message.

- **Chapter 7 —"Letters and Memorandums"**
  This chapter includes information and tips on the five types of letters used by most technical professionals: (1) letter of transmittal, (2) letter of instruction, (3) letter of follow-up, (4) letter of inquiry, and (5) letter of application. Also included is information on letter formats as well as some cautions on following popular letter-writing and envelope-addressing trends. This chapter also provides information on two basic types of memorandums — memorandums of transmittal and memorandums of information — and provides useful tips and information on how to develop attractive memorandums that will be read.

- **Chapter 8 —"Scientific and Technical Reports"**
  The technical report is one of the most common but most critical documents the technical professional will write. This chapter covers the important elements of a technical report and also information on related report standards published by the International Organization for Standardization (ISO) and the National Information Organization for Standardization (NISO). Although not all companies and organizations are required to adhere to these standards, the specified standards provide a comprehensive approach to report development that should help the technical professional develop quality in a consistent and accepted format. Helpful tips are also provided to assist in writing clear and concise reports. Examples of excellent technical report sections are included in this chapter.

- **Chapter 9 — "Procedures"**
  Procedures are other important technical documents that the technical professional will be required to develop or update. This chapter includes a discussion of document layout and development in a manner that is consistent with industry practices and standards. Tips are given on developing procedures, along with a flowchart on procedure development and updating. Several good procedure examples are provided.

- **Chapter 10 — "Proposals"**
  Today's technical professional will be called on to either develop or assist in the development of a proposal. In some instances the proposals will be submitted internally, while others will be submitted externally to generate additional company business. Although the respective request for proposals, or internal practices, may dictate the proposal format or layout, there are many valuable recommendations provided here for any proposal writing project. This chapter includes tips and information on how to

write proposals as well as a discussion on determining funding odds and how the proposal will be rated.

- **Chapter 11 — "Training Manuals"**
  As equipment and processes become increasingly more technical in nature, the need for training manuals likewise increases. Given the very nature of training manual development, this chapter provides information on two approaches: (1) recommendations on hiring external consultants to write the training manuals, and (2) recommendations on developing the training manuals on an in-house basis. In either case, the process of developing the manuals is essentially the same. This chapter includes a detailed discussion on the recommended steps for developing this important document from research to developing the manual to editing to validating its content. Included in this chapter are selected pages from a training manual that is an excellent example of a computer-produced manual that is both functional and attractive.

- **Chapter 12 — "Documentation, Duplication, and Distribution of the Technical Document"**
  Several of the most important, yet overlooked, steps in the document development process are the documentation, duplication, and distribution of the completed technical document. Because many technical professionals wait until the last moment before addressing the need to give credit where it is due, they have misplaced papers or simply forgotten that there are several undocumented references or citations. The result is that mandatory documentation elements are not included in the document; subsequently, suspicions of plagiarism may be levied by superiors, clients, and colleagues. Other problems associated with the final stages of document development have to do with duplication and distribution. Usually the technical professional has worked many hours to develop and produce a quality technical document that is sometimes turned over to others who may not be as interested in the quality. As a result, the technical document could be duplicated with inferior equipment and materials, making the final product not only unattractive, but also, in some instances, unreadable. Equally important, little consideration is given to how the document is to be transmitted to the intended reader. The result could be that the technical document does not reach the intended reader in a timely manner, which could mean loss of business or even loss of one's job.

  With the preceding factors in mind, this chapter includes critical tips and information on how to ensure that the document is documented, duplicated, collated, and bound in a professional manner. Information is provided to ensure that the technical document is received by the intended reader in a timely manner and in a way that helps protect proprietary information.

- **Chapter 13 — "The Technical Professional's Personal Library"**
  This chapter provides some thoughts and tips on why and how a technical professional needs to start a personal library of technical documents. A major emphasis of this chapter is on how to develop a library and maintain a broad collection of technical document samples that can be used to develop technical documents.

Final Comments

This is believed to be one of the most comprehensive books covering technical writing and the use of leading-edge technology to develop quality technical documents. Like any technical document, however, there is always room for improvement, and for this reason readers are encouraged to contact the author via e-mail with any comments, suggestions, or recommendations on how this book can be improved to better meet the needs of today's technical professional.

John H. White, Ed.D.
TechWrite_Inc@CompuServe.com

ACKNOWLEDGEMENTS

I have undertaken many large and daunting projects during my career, but none can match the hard work, fear, and sheer excitement I experienced in writing this book. This book has its roots in my short course on technical writing and in the encouragement of Ellen R. Kadin, former Editor of ASME Press. From there it mushroomed into a project that consumed more energy and resources than I would have ever imagined. Even so, I am eternally grateful to Ellen, the ASME Press, and the members of the American Society of Mechanical Engineers (ASME) for providing the opportunity for field practitioners such as myself to develop and publish technical books on a variety of topics that can help engineers, scientists, and technicians in their work.

Shortly before the manuscript was completed, John J. Corrigan was appointed the new Editor of ASME Press and immediately began working with me. Although John inherited me and my book from another editor, he nevertheless was supportive, firm, and fair — traits needed to be a successful editor dealing with temperamental authors. He was instrumental in having the final product published. Through the production stage I am particularly grateful to Colin McAteer who is extremely competent, patient, thorough, and who spent many hours in reviewing and coordinating production. Other ASME Press and ASME professionals who contributed to the publication of this book include Cynthia B. Clark, Philip Di Vietro, and Charles Beardsley. Their support and cooperation are greatly appreciated.

There were several individuals who helped with the development of this book. As I state in this book, having colleagues proof and edit technical documents is a critical step in the technical writing process. For the editing process I was fortunate to have two extremely competent colleagues who actually took pleasure in reviewing and picking apart my various drafts. Both Flo Sanderson and Patrice Stewart spent many hours reviewing the manuscript and provided comments and suggestions on how it could be improved. I incorporated all of their suggestions and believe that this is a better book thanks to their skill. A special thanks to the administration and staff of the Ralph Brown Draughon Library at Auburn University. A large portion of my research was conducted at the Draughon Library. During my many visits the administration and staff were extremely cooperative and supportive of this project. Angela Roberts, Librarian and Associate Dean of Academic Affairs for Academic Support at Alabama Southern Community College assisted in researching the names and addresses of the many editors and publishers that needed to be contacted in order to obtain written permission to use quotes and material from private or copyrighted sources. Obtaining the copyright permissions was one of the more difficult tasks associated with writing this book. Finally, I am grateful to Dr. John A. Johnson, President of Alabama Southern Community College, who encouraged me to write this book and had the vision to provide funding and support for a virtual library housed at Alabama Southern (e.g., on-line full text encyclopedias, dictionaries, etc.) that allowed me to conduct part of the book research near my home in rural Alabama.

One of the goals of this book is to provide engineers, scientists, and technicians with sample technical documents developed and used in actual industrial settings. Under existing national and international copyright laws, I had to obtain permission to use these materials in the book. Additionally, I had to obtain permission to use the tables, data, quotes, computer screen shots, graphics, and other copyrighted and non-copyrighted material used in the book. Obtaining these permissions was a tedious and sometimes frustrating endeavor and would not have been possible without the help from many professionals. Thanks to Russ Sirmon and Terry Wilkerson of the Alabama River Newsprint Company and Sloan R. "Corky" Fountain of VF Corporation, I obtained permission to use training manual samples and letter templates from two international companies located in my hometown of Monroeville, Alabama. Nancy Brunett, a chemical engineer with Advance Accessory Systems, assisted in obtaining permission to use several procedure samples. Nathaniel J. Leon, CEO of CDS, Inc., and Kenneth A. Crawford provided several samples of charts and memos. Helmut Thielsch, an internationally known engineer specializing in failure analysis, provided insights on technical report writing and provided sound advice on improving the chapter on technical report writing. Other individuals that contributed to the copyright permission efforts include John M. Solaski from Entergy Operations, Inc., Ted Brandhorst from the ERIC Processing and Reference Facility, and Mark Jeschke from the NASA Center for Aerospace Information. In addition, the following companies and agencies were extremely cooperative in providing permission to use their copyrighted material: Microsoft Corporation; Knight-Rider Information; CompuServe; Adobe Systems; U. S. Nuclear Regulatory Commission; International Organization for Standardization (ISO); and NISO Press.

Finally, I am grateful to my wife, Julie, and my mother, Shurley M. White, who were both extremely supportive and understanding about this project. There were many days and nights when I was late for meals or engagements or missed important events because I was working on this book. Without their support, I would not have finished this book.

OVERVIEW OF
TECHNICAL WRITING
IN THE WORKPLACE

Business and industry are now beginning to realize that a strong, positive correlation between good writing and document design can not only reduce its liability exposure but also save fiscal resources. In short, business and industry are beginning to understand that quality writing and design can be cost-effective.

Introduction

A few years ago engineers, scientists, and technicians applauded the arrival of the *information age* as an opportunity to reduce the many hours they spent writing and reading work-related documents and reports. Many truly believed that the arrival of technological gadgets such as the personal computer that ushered in the *information age* would be their salvation and give them the *paperless office.* Unfortunately, as Robert D. Ramsey (1993, 3) recently noted, "the computer which once promised to produce a 'paperless office' now seems to have conspired with the word processor, the laser printer and the fax machine to generate a never-ending demand for reports, memos, technical papers and correspondence of all kinds."[1] Many now find that possessing good technical writing skills is even more critical than before the arrival of the *information age.*

Today's engineers, scientists, and technicians find that although they received adequate, if not superior, education and training in their chosen fields, they are ill-equipped to conduct many simple writing tasks such as developing memos, reports, and manuals that document or communicate their work to others. Equally important, many find that because they lack essential communication skills their chances for advancement, or even job security, are severely limited.

This observation is best supported by the fact that today, engineers, scientists, and technicians, "spend over one-third of their work week writing, editing, or preparing oral reports" (Petroski, 1993, 419). The conclusion of one technical communications study found that the "managers were reading, the engineers were writing, and the supervisors were doing both." As Ramsey (1993), noted, the bottom line is that:

---

[1]Reprinted by permission of © National Research Bureau, P.O. Box 1, Burlington, IA, 52601.

In the modern work world, if you can't write a decent report, you've limited your opportunity for recognition, success and advancement. This is pretty scary for many professionals who feel that their expertise rests solely in working with data, numbers and technology — and not with the written word.[2]

Given the importance of technical professionals having the necessary writing skills for today's workplace, this book is designed to help engineers, scientists, and technicians (after this called technical professionals) to research, develop, and produce quality technical documents. Using narratives, illustrations, and actual industry examples, this book will provide the technical professional with practical and useful information that can be used on a daily basis in the workplace. This book follows a logical format and is organized in sequential order providing valuable information on the following eight basic types of technical documents:

1. Business letters
2. Electronic mail (e-mail)
3. Memorandums
4. Facsimile documents (faxes)
5. Scientific and technical reports
6. Procedures
7. Proposals
8. Training manuals

Additionally, the book will provide useful tips and information on research, planning the document, writing style, writing the technical document, avoiding common grammatical errors, document layout, desktop-publishing practices, and production of your work. In short, this is a comprehensive publication that covers all aspects of technical writing in today's workplace and has been developed under the premise that any competent technical professional can learn to write and produce quality technical documents.

The remaining portions of this chapter will review writing instruction provided in today's universities and colleges, the cost of technical writing problems in today's workplace, and the basic tools required for technical writing in today's technological workplace.

Technical Writing and Professional Preparation

During the past 10 years, colleges and universities have strived to provide better technical writing instruction to engineering, scientific, and technical program students. The primary motivation for these improvements was the demand from business and industry that graduates be competent in both written and oral communication. In response to these demands, most institutions of higher education attempt to integrate a high degree of writing into the respective course work. These efforts are also reflected in the *Criteria for Accrediting Programs in Engineering for Programs Evaluated During the 1996-1997 Accreditation Cycle (Criteria),* which was developed by the

---

[2]Reprinted by permission of © National Research Bureau, P.O. Box 1, Burlington, IA, 52601.

Engineering Accreditation Commission Accreditation Board for Engineering and Technology, Inc. (ABET). Under its enabling constitution, ABET (1996, 1) has two primary purposes:

1. Organize and carry out a comprehensive program of accreditation of pertinent programs leading to degrees, and assist academic institutions in planning their educational programs.
2. Promote the intellectual development of those interested in engineering and engineering-related professions, and provide technical assistance to agencies having engineering-related regulatory authority applicable to accreditation.

The specific criteria related to technical writing are provided in paragraph IV.C.3.i of the ABET *Criteria* (1996, 12):

> Competence in written communication in the English language is essential for the engineering graduate. Although specific course work requirements serve as a foundation for such competence, the development and enhancement of writing skills must be demonstrated through student work in engineering work and other courses. Oral communication skills in the English language must also be demonstrated within the curriculum by each engineering student.

Although ABET should be commended for identifying the importance of written communication skills, the criteria, as is, defer how such programs or courses will be conducted to the respective institution. The intentions of such criteria are that the engineering schools and technical program providers will be able to, or must, demonstrate that they have met this criteria during the periodic accreditation visits and reporting processes.

The result of this self-regulation is that an institution that understands the importance of its graduates having adequate, if not strong, technical writing skills will voluntarily provide comprehensive and curriculum-integrated technical writing in all aspects of the students' academic experiences before graduation. Additionally, because of this understanding and commitment, these same institutions are meeting the requirements specified in paragraph IV.C.3.i of the ABET Criteria.

On the other hand, there are institutions that feel their primary responsibility is to teach the engineering and technical subjects and they would rather leave the teaching of writing skills up to the faculty responsible for teaching general education courses. In a half-hearted attempt to meet the ABET criteria, however, some institutions may develop "technical writing" courses and have them taught by faculty from the English department. The only problem with this practice is that, according to Joseph (1989, 138) of the International Writing Institute, "most teachers of English are literature specialists. They do not want to teach composition, do not know how to teach it, and admit it privately."

Rather than focus on those institutions that fail to see the importance of providing quality technical writing across the engineering and technical curriculum, some insight will be provided on an institution that has, by all measures, exceeded the ABET criteria on written communication skills. During the past few years, the College of

Engineering at Clemson University has developed and implemented a comprehensive and integrated technical communications program known as *Effective Technical Communications* (ETC). This lighthouse program "integrates oral, written, and graphic communication exercises throughout the engineering curriculum" (Bennett 1990, 1384). As described by Bennett[1] et al. (1990, 1384–1385):

> This program is a supplement to courses in technical writing and/or public speaking required by the engineering disciplines in the college. It starts in the freshman year with motivational material on the importance of effective communication skills. In addition, freshmen receive an introduction to an analysis procedure which is applicable to almost every communication situation.
>
> First, students are taught to carefully determine their objectives. The second step includes the determination of the objectives, knowledge level, and needs of the intended audience. The analysis process also focuses on the selection of an appropriate medium and format. Finally, the importance of in-process evaluation, post-process evaluation, and modification of communications techniques is included.
>
> During the freshman year, students are also introduced to the ETC manual . . . As students become involved in their engineering courses, they are required to do communication exercises that are directly related to their studies. Written exercises include the preparation of memos, notes, letters, proposals, technical reports, and resumes. Oral exercises include short presentations on classroom material, special topics, and design projects.[3]

As mentioned in the program description, a critical element of the ETC program is the *ETC Manual,* which is used by students throughout their studies at Clemson University. This manual contains almost 200 pages of information and examples related to communication skills and serves as a valuable reference tool after the students complete their program of study and enter the work force.

An equally impressive element of the ETC initiative is its linkage with industry. One such linkage is the Clemson University Electric Power Research Association (CUEPRA) coordinated by the Electrical Engineering Department. The purpose of CUEPRA is to:

> Give visibility to electric power research and provide a working communications link between electric power companies and the academic community. In addition to providing opportunities for students to practice their communications skills, CUEPRA is instrumental in attracting high-quality students and faculty to the area of electrical power.
>
> CUEPRA undertakes research projects that are mutually beneficial to member power companies and the University. Funding from the power companies supports student research assistantships and undergraduate senior projects. The research projects are supervised by the faculty in the power system area. Graduate and undergraduate students make periodic

[3]Portions reprinted, with permission, from IEEE *Transactions on Power Systems,* Vol.5, #4, pp. 1384–1387, ©1990 IEEE.

presentations and write reports on their projects. Working on contemporary, practical problems of the electric power industry provides students with valuable, real world experience.[4] (Bennett et al. 1990, 1386–1387)

Both ABET and other institutions of higher education should take a closer look at the Clemson University ETC program and its successes. Hopefully, such a review would provide the necessary motivation for ABET, and its member institutions, to strengthen the written communications requirements of the *"Criteria for Accrediting Programs in Engineering in the United States."* This hope is best exemplified by an observation made by an engineering faculty member (who wishes to remain anonymous) of a leading college of engineering who was interviewed during the research stage of this publication. During the telephone interview, the faculty member said that "in the next 5 to 10 years almost 20 percent of the information we teach our engineering students will be obsolete and subsequently will not be a usable skill. In contrast, technical writing is a skill that we can provide our students which *can be used their entire career* [italics added]" (Doe, 1994).

In the anonymous professor's comments on the value of providing students with these skills before they graduate and enter the work force, the relationship between an engineer's technical writing skills and professional advancement is an empirically documented fact. For example, the American Society for Engineering Education (ASEE) conducted a study of 4,057 "seasoned" engineers to help determine which academic subjects were most critical to an engineer's career. As shown in Table 1, technical writing was identified as the number two most needed skill (Olsen and Huckin 1991, 4).

Table1 Subjects Most Needed for Engineering Careers in Industry

Subject[1]	Rank
Management practices	1
Technical writing	2
Probability and statistics	3
Public speaking	4
Creative thinking	5
Working with individuals	6
Working with groups	7
Speed reading	8
Talking with people	9
Business practices (marketing, finance, economics)	10

Source: Olsen and Huckin, *Technical Writing and Professional Communication, 2d ed. (New York: McGraw-Hill, 1991),* 5. Reprinted with permission of the McGraw-Hill Companies.

[1]The 4,057 engineers surveyed included all engineering fields ranging from mechanical engineering to civil engineering. Given this broad range of engineering fields, technical or engineering skills for a particular group of engineers (e.g., subjects related to heat transfer, materials, fluids, etc.) may have ranked higher as a subgroup of engineers (mechanical engineers) but not for all surveyed engineers. As such, technical writing was a most needed "academic" subject (second only to management) identified by a majority of the engineers surveyed. Reprinted by Permission from McGraw-Hill, Inc.

[4]Portions reprinted, with permission, from IEEE *Transactions on Power Systems,* Vol.5, Nov 4, pp. 1384–1387, ©1990 IEEE.

Finally, another related study conducted by the University of California at Berkeley illustrates the relationship between technical writing skills and career advancement. The study included participation of 595 engineering alumni, of whom, when asked if "writing skills had aided their advancement, 73% answered yes" (Olsen and Huckin 1991, 4). Equally important, "almost all (95%) said that they consider writing ability in making hiring or promotion decisions." (Olsen and Huckin, 1991,4)

Reasons for Technical Writing

Beyond the self-centered reasons (e.g., employment, job security, promotion, etc.) for having strong technical writing skills, there are other equally important reasons. Independently, these reasons are compelling enough, but together they are a graphic reminder of just how important the art of communication is to the success and survival of any organization or company. There are several such reasons, including:

- *Legal* — In today's highly litigious society it is important to properly document selected activities and initiatives to reduce the liability exposure of the professional and company. The legal documentation can show that the individual was prudent and took all proper actions and procedures. In addition, the legal documentation can "refresh one's memory" on what did or did not take place. Often, it takes years for legal cases, such as a wrongful death case in an industry accident to actually reach the deposition or court stages. Given this long period, individuals tend to forget all of the facts. Worse yet, the facts become distorted. Many legal cases have been won by the defendants because they had meticulous and clear documentation on related events which clearly showed that the defendants took all proper measures and were prudent in the related decision-making process. Similarly, many defendants lost their legal cases — not because they did not follow proper procedure or were not prudent, but because they lacked the clear documentation showing such practices. The result has cost industry billions of dollars and has cost otherwise competent professionals their jobs and careers. It is beyond dispute that given this highly litigious society, a company will not rush to hire a professional who has cost his former employer millions of dollars due to poor documentation related to a legal case or suit — that would not be the prudent thing for management to do.
- *Safety and Welfare of Employees, Customers, and the Public* — The importance of technical writing is perhaps no more apparent than when ensuring the safety and welfare of an organization's employees, customers, and public. An ideal example would be a nuclear power plant that when operated properly can provide a tremendous benefit and important service to society. Due to the obvious sophistication and complexity of such a facility, even the most minor operation details must be documented

through written procedures. These procedures must be comprehensive and clearly written so that there can be no misunderstanding on the intent and meaning. Equally important, assigned engineers, scientists, and technicians are routinely required to update or write new operation procedures as the facility's equipment is updated or if procedures must be clarified due to incidents or changes in related regulations, codes, or standards — all with the intention of protecting the safety and welfare of all concerned.

- *Transfer of Knowledge Gained From Research and Development Activities* — Due to rapid technological change, the engineering and scientific knowledge base is increasing at a remarkable pace. This necessarily means that almost all technical professionals will be required, at some point in their career, to participate in research and development activities. An allied responsibility will be to develop or help in the development of a research and development document to share gained knowledge with other technical professionals. Such practice has been the responsibility of technical professional for decades. In fact, as John Alexander Low Waddell stated in a vocational guidebook to the engineering profession in the early 1900s:

> Whenever an engineer learns something new in technics, it is his bounden[sic] duty to put it in writing and see that it is published where it will reach the eyes of his conferees and be always available to them. It is absolutely a crime for any man to die possessed of useful knowledge in which nobody shares.(Petroski 1993, 421)

Waddell also believed in the importance of communicating engineering to the public:

> Every established practicing engineer should also, on occasions, contribute to magazines, or the daily press timely descriptions, discussions, or statements of local or important engineering projects or constructions, clearly showing the layman the vital features, advantages, disadvantages, and probable or possible results.

- *Manuals or Instructions on the Proper Use of Equipment* — Once a company has completed the development or upgrade of a piece of equipment or even software, documentation must be developed on the proper and safe operation. Increasingly, organizations will base their purchasing decision on a comprehensive review of a given product. One of the many elements reviewed will be the accuracy and "user-friendliness" of the operating or user manuals. Today's companies are keenly aware of the fact that poorly written user manuals can cost thousands if not millions of dollars in employee learning time. Likewise, a comprehensive and clearly written manual can reduce product liability claims — this is especially true if the manuals or instructions include information on the proper care and use of said equipment as well as the "don'ts."

- *Development of Industry Codes and Standards* — Technical professionals are regularly called upon to help develop industry codes and standards and respond to codes and standards implemented or under consideration for implementation. In all instances, clear and concise technical writing is required. Who would be in a better position to provide such critical input than the engineer in the field who helped develop or works with the related equipment or process on a daily basis? Again, clear and concise technical writing or related response documents will convey the intended message or concern. The result may be a change in the respective code or standard that ultimately saves industry and the consumers millions of dollars, not to mention the safety and welfare of employees and public.

- *Generate Additional Business or Income* — Technical professionals are regularly called upon to develop proposals for their employers in response to requests or invitations to bid. Unfortunately, as many technical professionals have learned, providing a bid proposal with the lowest cost is not the only factor that will be considered in awarding a bid. More often than not, other important factors will be considered in such decisions, covering a full spectrum from the quality of the project plan to the quality of the project management plan. As will be shown in later sections, sometimes the clarity, presentation format, and even packaging can be extremely important factors in determining which contractor is to be awarded the contract.

- *Day-to-Day Communications With Colleagues and Customers* — Today's society is an "instant" society. Coupled with this fact, business and industry are going through an industry phase called "downsizing." The result of downsizing is that there are fewer employees left in an organization to do the work that was previously done by many more employees. In turn, it is much more difficult to have face-to-face meetings, not to mention almost impossible to reach colleagues and customers by telephone. To fill this resulting communications gap, technical professionals must now rely on new instant forms of communications methods, which, as a minimum, include faxes and e-mail. In such situations the importance of clear and concise communications cannot be overemphasized. Unlike a face-to-face meeting where one can rely on nonverbal communications (e.g., facial expressions and other types of body language, etc.) and immediate verbal communications to verify our intended meaning, one must be very clear in the highly technological communications to reduce or eliminate any possible miscommunication of the intended message. Keep in mind that unclear written communications have been misunderstood and have cost industry billions of dollars.

- *Persuade Colleagues, Employers, and Customers* — Within any organization or society there will be conflict or disagreements on ideas and resolutions to problems (sometimes called "opportunities"). Ultimately, in an industrial or technical organization it will be left to the technical professional to draft a position paper, memorandum, or letter to help persuade the target group or individual make a decision on the proper action. Many times over, competent engineers are unable to develop a clearly written document to help persuade others on a given course of action. The difference between writing a clear and persuasive document may very well mean the difference between the life or death of a project, which, in turn, can be the difference between being gainfully employed and not being employed.

The Cost of Technical Writing Problems

Unfortunately, there is no clearing house or organization that keeps track of just how much money poorly designed and written documents cost business and industry annually. Some writing experts have, however, compiled accounts or case studies on just the financial impact poor technical writing can have on a company. For example the videotape *Power Writing* (1990), by the editors of *Communications Briefing,* illustrates how one poorly written letter to almost 60,000 customers had to be clarified through an additional letter to each customer. The additional postage alone for this clarification must have cost the company at least $15,000. In a broader example, Schriver (1993, 249) of Carnegie Mellon University states that "a growing number of product liability cases on record indicated that poor-quality documents are costing manufacturers hundreds of thousands of dollars."[5] Schriver adds that according to P. S. Helyar, the courts are now holding manufacturers accountable for not providing quality product documentation, which includes better written instructions and warnings. Specifically, Helyar states:

> Since 1974, the number of product liability actions in federal courts has increased eight-fold. In many of these actions, plaintiffs claimed that inadequate directions or warnings caused their personal injury or loss. This proliferation is not due to increased mediocrity in documentation. It has been caused by the courts' growing tendency to impose liability on manufacturers who, for pennies per product, could improve directions and better warn against unexpected dangers in their wares.[5] (Schriver 1993, 249)

According to Schriver, there are many cases where companies have been sued because of "factors related to poor-quality documentation." One such case is a "settlement in 1986 in excess of $945,00 paid by John Deere & Company to a plaintiff who suffered severe and permanent injuries when a crawler loader he was operating slid on an embankment and rolled over on top of him. Deere & Company had failed to include

---

[5]Used with permission fron *Technical Communication*, published by the Society for Technical Communication, Arlington, VA.

appropriate warning and instructions in the operator's manual"[6] (Schriver 1993, 249).

Business and industry are now beginning to realize that a strong, positive correlation between good writing and document design can not only reduce its liability exposure, but also save fiscal resources. In short, business and industry are beginning to understand that quality writing and design can be cost-effective. This statement is best supported by two examples on the savings realized by companies and government organizations that have rewritten and designed forms, reports, and other documents:

- The Motorola Corporation Finance Department has substantially improved its operation after a quality movement. In fact, they won the Baldrige[7] in 1988. They now close their books in four days, down from 12 in 1987. Changes such as clearer directions on forms and an easy-to-use format for computer screens have helped streamline the process — and save $20,000,000 a year[6] (Schiver 1993, 250).
- A technical publications group at AT&T reports that after implementing a quality program between 1987 and 1988 that focused on streamlining the process of technical documentation, they reduced the cost of documentation by 53%, reduced documentation production time by 59%, and increased the number of projects individual writers were able to complete by 45%[6] (Schiver 1993, 251).

Take Advantage of Problems as an Opportunity to Hone Technical Writing Skills and Enhance Career Advancement Opportunities

Technical writing has "often been described as a form of problem-solving" (Eklundh and Sjöholm 1991, 747). And in the engineering, scientific, and technical community, problem-solving is the very "soul" of the profession. Obviously, almost every organization that employs or uses the services of technical professionals will have many problem-solving opportunities. The shrewd technical professional can take advantage of these opportunities by volunteering to help in the documentation and reporting of these problem-solving activities. Usually colleagues will gladly defer the tedious technical writing tasks to a volunteering technical professional. The volunteering technical professional can profit in several ways:

- *Hone or Improve Technical Writing Skills* — First, one of the best ways to improve your writing skills is to write, write, and do some more writing until the writing process and mechanics become second nature. Technical writing is a skill and like any other skill— the more you practice, the better you become. Technical writing is also an acquired taste, similar to coffee, which at first tastes a little on the bitter side. Over time, however, the more coffee you drink, the more you like and crave its aroma and taste. Thus,

---

[6]Used with permission from Technical Communication, published by the Society for Technical Communication, Arlington, VA.

[7]The Baldrige National Quality Award (Baldrige) was named after the late Malcolm Baldrige, Secretary of Commerce under President Reagan. The Baldrige was first awarded in 1987 (Schriver 1993, 241).

although you may presently find technical writing a little on the bitter side, the more practice you have, the better it will feel to the point that you will look forward to those moments that you can sit at your desk or computer and write.

- *Improve Reasoning and Decision-Making Skills* — As technical writing is a problem-solving process, "learning how to write well is an opportunity to learn how to deliberate, how to bring principles and concrete facts to bear on a situation that requires decision and action"[8] (Johnson 1993, 398).

- *Learn the Subject Matter* — The technical writing process is not unlike teaching. As seasoned educators know, the best way to learn a subject is to teach it. The public will be shocked to know how often educators teach a subject that only a few weeks or days before they knew very little about. In preparing to teach a class, there are several steps that an educator will go through to learn or become familiar with a given subject. This process is also used in technical writing. In fact, because most technical professionals lack adequate technical writing skills, many companies have to hire technical writing teams to come into their organizations to develop and write procedures and training manuals. What is most interesting about using professional technical writers is that more often than not, these professionals have a teaching background but have very little knowledge on the specific topics they are hired to write about. What these professionals do possess, however, is good technical writing skills and the ability to learn any subject (enough to write about it, that is) in an extremely short period of time. In short, through experience these professionals have learned to shorten their "learning curve" on almost every conceivable subject. Of course, they must rely on the field engineer to give them the technical information and to review the draft and completed technical document.

  Like the professional technical writer, the technical professional can take advantage of the technical writing process to help learn new subject matter. The result will provide the technical professional with more work-related knowledge that can only help in future endeavors and can help foster job security and even professional advancement opportunities.

- *Provide Visibility to Management* — Given today's busy workplace, upper management rarely has opportunities to talk with and get to know all of the company's employees. Volunteering for technical writing assignments can thus provide the technical professional additional exposure to and recognition of upper management. This exposure can, if nothing else, imprint your name and willingness in the minds of upper management and can prove to be beneficial in assignments of other, perhaps more exciting and important, projects and promotions.

[8]Reprinted by kind permission from Kluwer Academic Publishers.

Technical Writing in the Information Age and Tools of the Technical Writer

Fortunately, we are living in the *information age,* which means, among other things, that there are new and emerging technologies to develop and transmit information. No doubt the personal computer is one such technology that has already had a tremendous positive impact on society and how work is done. With the personal computer, one can analyze data at a desk; a few years ago, that could only be done at a major research center that had an expensive mainframe computer system. Also, one can develop sophisticated-looking technical documents and reports required for today's jobs without the costs and delays previously associated with going to an outside print shop. Although, as discussed earlier, the arrival of the personal computer will not necessarily free workers from work or do their work, it can certainly help an ividual become more creative and do a task that previously took many workers to complete.

One of the basic tools required for technical writing is the word processor, which, when used with a late-model computer, can help the technical professional in developing quality documents that are also professional looking. With today's "high end" word processing packages (e.g., WordPerfect or Microsoft Word for Windows) the technical professional can develop and edit the technical document at the computer keyboard, thus saving a great deal of time "used for planning the text and beginning of the writing process" (Eklundh and Sjöholm 1991, 747). Ideally, using the computer directly in the writing process (composing at the keyboard) has several advantages and is becoming a common practice by more technical professionals. Beyond saving time in the writing preparation stage, it also gives the writer full control of the writing and editing process and even in the production of the document. Also, due to the industry downsizing trend, many do not have the same secretarial support as in the past — thus it is left up to the technical professional to not only develop the document, but also to be responsible for its word processing as well. As shown by Eklundh and Sjöholm, there is some "folklore" about using computers in the writing process. "For instance, it has been claimed that the computer makes writers write sloppily or speech-like, and this would make their texts less readable" (Eklundh and Sjöholm 1991, 724). On the other hand, there is a positive note: "many writers report that the flexibility of the computer makes it possible for them to write down early ideas in a more spontaneous way, and thus to get more of their thoughts into print." (Eklundh and Sjöholm 1991, 724)

Ultimately, whether to use the computer or not use the computer directly in the writing process will be a personal choice. Either way, a computer will be used to generate the text or document. The only negative side of using the computer rather than the typewriter is that the word processor provides powerful editing capabilities, which can lead to editing a document to death. As most writers will eventually learn, there comes a point of diminishing return; no matter how much "fine-tuning" is done, the document does not get better, and may get worse. Given this axiom, it is important for the technical professional to keep track of the amount of time taken to develop and edit the technical document, and there must be a logical point of actual completion of the document.

Coupled with the word processing of the document is the use of today's sophisticated desktop-publishing systems. The best working definition of a desktop-publishing system is that the programs provide for the full integration of text and graphics in one software package that also provides for the complete control of all elements from color to kerning to sizing of the printed document. Additionally, a desktop-publishing system allows the user to produce "camera-ready" documents that can be transmitted to a print shop for production with exact colors or gray scaling required, or to the copy room for simple duplication.

Today, the distinction between word-processing and desktop-publishing programs is not as clear as it once was. That is, many of today's high-end word processors are marketed as desktop-publishing packages, as they can integrate text and graphics. To a purist, however, such claims are tantamount to saying a motorcycle and an automobile are the same — after all, they're both capable of traveling on the highway at high speeds and can transport people. Such claims may be true, but how much longer will it take for a motorcycle rather than the automobile to transport six people to a location 100 miles away? And, what kind of shape will these six people be in when they finally arrive by the motorcycle as compared to being transported in the automobile?

In addition to a personal computer, word-processing program, and desktop-publishing program, there are several other tools that, as a minimum, every technical professional should have before starting the technical writing process, including:

- *Laser Printer* — Having a laser printer to produce technical documents is a given in today's highly technological workplace. The cost gap between a dot-matrix printer and a laser printer has decreased to the point that it is inconceivable why dot-matrix printers would be used in the production of technical documents.

- *Modem* — A modem will provide the capability to conduct international and comprehensive research in the comfort of your home or office. Ideally, purchase the highest speed, latest model modem on the market. Although these later modems cost a few dollars more than the older models, you will ultimately save on long-distance costs and on-line connection charges due to the faster data speeds. Additionally, most of the higher speed modems on the market today are fax/modems that have the capability to receive faxes through the personal computer. This will be useful during the technical document research stage and will be discussed in Chapter 2.

- *Latest Version of Grammatik*™ — This sophisticated grammar checker and spell checker interfaces with today's high-end word processors and desktop-publishing software packages. The greatest benefit of Grammatik is the grammar-checker engine that generates usable alternatives for words and phrases and helps ensure consistency in tense and usage, producing shorter and more readable technical documents. Also, other package elements include readability and document statistics that can help technical writers develop and fine-tune technical documents to meet the reading and comprehension levels of their target audience.

- *The Chicago Manual of Style* — The latest edition of the reference book is essential for any technical writer. This industry standard covers every conceivable writing topic from title pages to reference notes. It is the standard choice of hundreds of publishers and is used by almost every professional writer.
- *The Elements of Style* — Written by Strunk and White (1972), this guide was originally published in 1959 and is still considered a classic and "one of the best books on the fundamentals of writing."

Naturally, depending on individual financial situations and style, modifications will be made to the list of tools of the technical writer. The main point, however, is that in today's highly technological and globally competitive world, these are the necessary tools to properly research, develop, and produce quality technical documents.

RESEARCH

"The world is changing rapidly, and businesses are no exception. To keep up with — and surpass — competition, you need information faster. A wider variety of people within your organization need information at a moment's notice, and some want to search it themselves. This, coupled with widespread corporate downsizing, is forcing you to do more with fewer resources" (Tierney 1995, 3).

CEO P. Tierney
Knight-Ridder Information, Inc.

Introduction

The first logical step in developing a technical document is to think about what you want to write and what information is available. One of the biggest mistakes made in the technical writing process is to start writing with the idea that you can conduct research as needed to fill the gaps and holes. Worse yet, many writers start writing their documents thinking they know all they need to know. Except for a one-line letter or memo, this can be a fatal assumption. Operating under the assumption that you already know all you need to know can lead to developing a shallow document that fails to include pertinent information or findings and could lead to incorrect conclusions. In some situations, incorrect conclusions can be professionally embarrassing and even cost you your job.

To ensure that you do have all the available information on your writing project and can then develop a document that has depth, is timely, and can be of value in the problem-solving process, you must first conduct "comprehensive research." Comprehensive research can be undertaken in an orderly process that encompasses six sources, including:

1. What you know
2. Company information
3. Public library systems

4. College and university library systems
5. Internet
6. Commercially available computer databases (which can be extremely costly)

The remaining portions of this chapter provide information, including the advantages and disadvantages, on each of the six research sources.

What You Know

The plan for the development of a technical document is to start early. Waiting until the last minute not only means that you will rush through developing your document, which means you can overlook or omit critically important information, but also that you may run into some unforeseen problems (e.g., gaps in available data, missing parts of other reports, missing important project notes, or information a colleague has failed to provide you with, etc.), and additional time will be required to resolve the problem. An early start on the project can help you avoid some of these unforeseen, but common, problems that most technical professionals experience once they begin drafting the document. The rule for starting early applies to the development of every type of technical document — from the simple memorandum to the complicated training manual.

Like most projects, one of the most critical points is starting the project. Simple logic says if you never start, then you can never finish. Traditionally, most people have trouble at two critical points: they procrastinate when starting a project and face similar problems completing it in a productive and timely manner (more on finishing will be provided in a later chapter). One of the best ways to help overcome these procrastination tendencies in beginning the project is to start early by getting organized and gathering the information you already have. People organize their material differently, whether in file folders that are color-coded to differentiate between certain types of support information or documentation or the more time-consuming process of using computer databases to index all materials. Naturally, how you organize your materials is more a matter of style and will also depend on the scope of the project and the amount of information gathered. The more information you have, the more you may need computer indexing to ensure that you do not omit any important information or data.

There are at least four types of information that you have that should be collected and reviewed, such as the following:

1. Letters and memorandums
2. Reports
3. Personal notes and logs
4. Personal files

Company Information

After you have collected and centralized your personal project information, the next step is to sift through company information on the project. Again, this phase should be initiated as early as possible. Usually there are at least seven types of company information that can be collected and used, including:

1. Source documents, including related manuals, engineering reports, drawings, and test specifications
2. Letters and memorandums
3. Reports
4. Marketing plans
5. Beta test reports
6. Safety reports
7. Purchase orders and invoices

These documents generally contain a wealth of information that can be of great help when you are developing your document. Unfortunately, you may find that not everyone in your company will cooperate and provide you with this information. Also, the larger the company or organization, the more trouble you will have collecting this critical information. Issues such as "turf" and excuses such as "confidentiality" or "security" concerns will be generated. The more creative and experienced "blockers" will assure you that you can have all of the requested information when it is "found."

These internal collection problems are more common than most professionals would like to admit. The best way to deal with these problems is to first collect all the information and data that you can without incident. Next, consult your supervisor about what actions to take with company personnel who are not cooperating. In fact, the best course of action is to ask your supervisor to intercede and obtain the requested information so that you can proceed to the next step in collecting information and data.

Public Library Systems

Since the days of Benjamin Franklin, who is considered the father of today's public library system, public libraries have become a cost-effective means of providing citizens with equal access to all forms of materials, including books, books on tape, records, compact disks, newspapers, journals, and magazines — all free to the public through the support of tax dollars and donations. Currently there are approximately "15,000 public libraries in the United States" supported primarily through local, state, and even federal funds (*The World Book Encyclopedia* 1995).

While the public library is a commendable institution and provides a great service to local citizens, elementary and high-school students, and in some instances college students, the value of the public library to help in the development of your technical documents will be directly dependent on its location, number of holdings, and number of journals and serials. The main branches of such systems in New York, Detroit, Chicago, and other large metropolitan areas will surely be a comprehensive research source that could very well contain all of the information you will need. Smaller urban and residential community libraries generally would not contain the scope of materials needed. Some smaller community libraries will have the same resources as those in larger urban areas, but they will simply be the exception rather than the norm. In short, the community public libraries were not designed to be all things to all people but developed to meet the needs of the local citizens. Conducting sophisticated research on technical, engineering, and scientific topics is generally not their mission.

To make a quick determination whether your local library contains the research resources you require, visit the main branch and talk with the director or head librarian and explain your needs. By nature, librarians are extremely helpful and provide guidance. Unlike most other professionals, librarians will not hesitate to send you to another area library that may better meet your needs.

College and University Libraries

Most professionals remember the many hours spent in the library doing research during their college years. For some these experiences were enjoyable, while for others it was more like work than a productive learning process. Now these same individuals are in the workforce and realize that research is important in the career of the technical professional. Equally important, some of these professionals are led to believe that today's libraries are being replaced by the "information superhighway" and are doomed to extinction as was the dinosaur. However, college and university librarians have carefully and methodically brought leading-edge computer and information technology into the library to provide better and more efficient services to their patrons; lobbied their college presidents and legislators for more funding to keep up with technology; maintained a critical balance between the number of traditional library holdings (books and serials in the stacks) and computer databases; and are doing all they can to preserve older and rare books for posterity.

In short, the librarians of the nation's research colleges and universities are actively embracing technology for the betterment of services and are also ensuring that their facilities will be around for many years. The continued preservation of college and university libraries is welcomed by many researchers who have found that there is no true replacement (not in the near future, at any rate) for library-based research. The key to the utilization of library-based research will be time and the availability of a major research library.

Unlike the commercial document fulfillment centers, on-line services, and other commercial services, the research university is not, and should not be, a profit center. These institutions will consume large amounts of tax dollars and private support to continue to operate and upgrade services as technology changes. They are the holders of knowledge and history and should never be forced into positions of "profit." This does not mean that the research college and university libraries do not contribute to the net growth of our society and nation. To the contrary, college students and employees of business and industry nationally take full advantage of the free services offered by these critical institutions daily. In fact, almost every research and development, marketing, production, and legal project generally starts with library-based research. Given this high usage, it is safe to say that if the doors of these great institutions were closed, it would take only a short time to bring the nation's economy to a complete halt due to the lack of critical information.

Given the importance of the research college and university libraries and today's expanding information commerce, most research projects will be a combination of activities. Time permitting, most research projects will begin in the library where both books and scholarly journals can be easily identified through sophisticated Boolean computer searches (key-word searches using operands such as "and," etc.). The library-based research will be coupled with the use of on-line and commercial document fulfillment services for those sources that are not within the respective library holdings or computer indexes. The combination of library-based research, on-line research, and commercial databases will help ensure that the professional has attempted to obtain information from every available source — both public and proprietary information.

The key to success during the first library-based research stage is to go to a library of sufficient size in terms of library volume holdings and availability of scholarly journals and serials in the field of research. Thus a mechanical engineer would ideally go to a nearby state university that had a reputable engineering program to conduct research.

With a few notable exceptions (e.g., Harvard and Yale), most of the major research colleges and universities are state institutions supported by a combination of public and private funding. The top 50 major college and university libraries, and their respective holdings (volumes, journals, serials, microforms, etc.) are provided in Table 2.A.

Hopefully, one of these libraries will be nearby and you will not have to travel great distances to do your research. Also, just because the college and library near you are not on the above list does not mean that they cannot meet your needs. Rather, the list is provided to give you an idea of the relative size (in terms of volume and microfilm holdings) to make some judgments about other libraries that you may consider visiting. Prior to and during your library-based research activities, there are several steps that should be taken to ensure that your library visits are fruitful and efficient:

- Call the head librarian to get an overview of services offered, hours of operation, and cost associated with duplication of articles. Also, make certain that the library participates in interloan of materials with other research college and university libraries. Finally, ask questions about the availability of books, journals, and serials in your field of interest. Related questions should be on what indexes and full-text journals are available on computers.

Table 2.A Holdings of Major U.S. College and University Libraries

Institution	Volumes	Microforms
Harvard	12,394,894	6,335,484
Yale	9,173,981	4,020,003
U of Illinois-Urbana	8,096,040	3,897,360
U of Calif.- Berkeley	7,854,630	4,666,786
U of Texas	6,680,406	4,779,137
U of Michigan	6,598,574	4,495,426
Columbia	6,262,162	4,460,914
U of Calif. - Los Angeles	6,247,320	6,377,470
Stanford	6,127,388	4,119,839
Cornell	5,468,870	5,833,564
U of Chicago	5,448,621	1,939,074
U of Wisconsin	5,317,380	3,755,241
Indiana	5,264,138	2,971,921
U of Washington	5,163,302	5,730,164
Princeton	4,965,358	2,813,879
U of Minnesota	4,908,982	3,936,564
Ohio State	4,603,310	3,549,712
Duke	4,134,361	1,719,587
North Carolina	3,956,238	3,462,277
U of Arizona	3,915,913	4,326,629
U of Pennsylvania	3,844,414	2,626,937
U of Virginia	3,842,344	4,871,904
Northwestern	3,607,533	2,539,884
Rutgers	3,367,663	4,314,783
Pennsylvania State	3,318,118	3,333,266
Iowa	3,253,141	4,873,174

- Ask if the library automation system [on-line public access catalog (OPAC), database indexes, and even on-line full-text articles of some holdings] has dial-up or Internet capabilities. More universities are providing such services for a small charge. Having the dial-up access will mean that you can conduct a large portion of the research at your office or home. In fact, if you are looking for only a few articles on the research topic, having access to the indexes through the dial-up process will identify the article that can be ordered from the library. Most libraries will fax such articles in emergency situations. Of course, you can always purchase the articles from UMI (discussed in a later section of this chapter) or other document fulfillment center. In most instances, the library will have an OPAC for patron use in the library. OPAC is a large database that is a computer version of the old card catalog system. This computer card catalog provides patrons with the capability to conduct searches of library books (holdings) based on author, subject, title, key-word, and even the call number (see Figure 2.A). Once a patron has located the desired publication, a screen that looks similar to a card from the old card catalog

TABLE 2.A Continued

Institution	Volumes	Microforms
New York	3,224,145	3,074,824
U of Georgia	3,048,491	4,865,800
U of Kansas	3,043,964	2,510,217
U of Pittsburgh	3,041,139	2,811,102
U of Florida	2,974,962	4,876,516
Johns Hopkins	2,961,160	1,842,302
Michigan State	2,860,874	4,610,166
Arizona State	2,826,679	4,410,042
Rochester	2,774,892	3,661,525
U of Southern Calif.	2,764,865	3,765,719
SUNY-Buffalo	2,724,222	4,093,112
Washington U-St. Louis	2,703,584	2,469,803
Wayne State	2,667,088	3,271,271
Louisiana State	2,654,485	3,923,432
U of Hawaii	2,651,257	5,242,164
U of California-Davis	2,588,728	3,082,956
U of Missouri	2,579,253	5,012,692
Brown	2,551,187	1,236,258
South Carolina	2,526,408	3,777,699
U of Massachusetts	2,511,558	1,911,953
U of Kentucky	2,459,497	4,861,232
U of Colorado	2,427,603	4,524,983
U of Oklahoma	2,381,304	3,216,398
U of Connecticut	2,372,350	3,228,043

Source: Research Libraries. The *1994 Information Please Almanac*
CompuServe, Columbus, Ohio (1995).

is shown (see Figure 2.B). An added feature of OPAC is an indication if the respective publication is available for checkout.

- Ask if the library has full-text articles of leading journals on-line. The more progressive libraries will take advantage of the new technologies that enable development of systems that provide full-text articles on-line for searches and printouts of articles from professional and popular journals and magazines. The Information Access Company, for example, offers the *Academic ASAP*, an on-line service that enables users to locate and print full-text articles in seconds. Updated monthly, this CD-ROM database provides convenient access to more than 220 of the scholarly and general interest periodicals most requested by academic librarians, students, faculty, and researchers. This database covers a wide range of topics:

  — Humanities
  — Book Reviews

  — Communication Studies
  — Current Issues

— The Arts
— Science and Technology
— National and
   International News
— Public Affairs and
   Public Policy

— Social Studies
— Women's Studies
— Environmental
   Studies

A sample search screen of the *Academic ASAP* is provided in Figures 2.C, 2.D, and 2.E. Also, Figures 2.F.1 and 2.F.2 provide a copy of an article printed from the *Academic ASAP*. The research power of these full-text systems is extraordinary and will save the researcher hundreds of hours in terms of searching, retrieving, and copying journal articles.

- Ask to have a complete tour of the library before you begin your research. The tour should be conducted by a staff member who knows the complete library system — holdings of journals, serials, and resources available on the automated library system. Do not settle for a work-study student or a new staff member who will not be familiar with all services available.
- Talk with library staff as you are conducting your research. Find out who really knows about the library and ask these key professionals when you cannot locate materials in the stacks.
- Check whether the copy machines require change or whether the library sells voucher cards. More libraries are using voucher cards that you purchase in quantities of $1, $5, $10, and even $20 for use in the copy machines. These cards can save much time and money, and you will not have to carry several extra pounds of change.

**Figure 2.A Sample Main Menu Screen
From the On-line Public Access Catalog (OPAC) at
Alabama Southern Community College in Monroeville, AL**

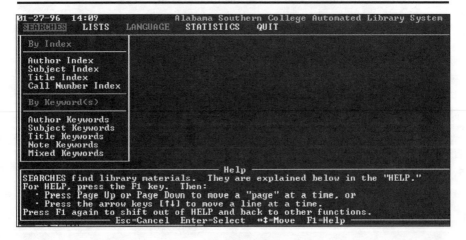

Figure 2.B Sample Book Holding Screen From the On-line Public Access Catalog (OPAC) at Alabama Southern Community College in Monroeville, AL

```
01-27-96  14:29      Subject entered: Chemical laboratories--Safety measures.
Full Info.   Availability    Display format    Main Menu
                         Card 1 of 1
T'ville     Handbook of laboratory safety / A. Keith Furr, Chemical Rubber
Ref         Company. - 3rd. - Boca Raton, FL : CRC Press, 1990.
OD          704p.
51
.H27        $100.00  4-93.
1989        T'ville Ref OD 51 .H27 1989.
            RSN 00025261.

            1. CHEMICAL LABORATORIES--SAFETY MEASURES.  I. Furr, A. Keith
            II. Chemical Rubber Company.  III. Title: CRC handbook of
            laboratory safety.
            No copies of this title have been recorded for checkout
                                 Help
To see information about library copies of the selected title,
press Enter.
If the message "No copies of this title have been recorded for checkout"
is displayed, the library does not have any copies of the title to
circulate.
            Esc=Cancel   Enter=Select   ↔↕=Move   F1=Help
```

- The library staff of many larger research libraries will also conduct comprehensive searches on commercial databases. Such services are available for special requests where the research-connected time and staff member's time is charged to the respective patron. If you are not really experienced with computers or do not have much time, use of these services can be a tremendous help. A word of caution: be sure that you set some cost limits with the library and make certain that the staff member conducting the custom searches is an experienced researcher who has experience with the respective databases; failure to obtain these assurances can cost a great deal of time. More important, relying on the skills of an inexperienced researcher could mean that some valuable data or information is overlooked.

- Even if the library does not have full-text articles on-line, most research libraries have created large databases of indexes of journals, magazines, and serials of library holdings. These systems, sometimes called automated library systems, generally permit computer searchers by author, title, subject, and key-words and will save the researcher a considerable amount of time. When using the library's automated system to conduct a search of indexes, make certain that you are accessing the correct databases. You would not, for example, want to conduct a search through the "general periodical indexes" when you actually want to conduct a search through the engineering or technical indexes. Depending on the type of library automation system used, the indexes could all be merged into one database, accessible on the same computer workstation but through a separate menu, or even on a separate computer workstation at another

Figure 2.C Sample Screen From the Academic ASAP System at Alabama Southern Community College in Monroeville, AL

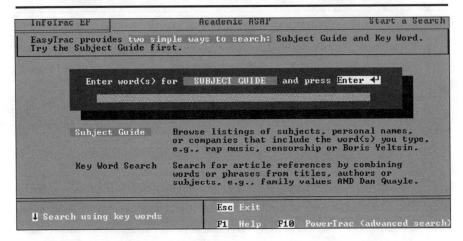

location in the library. Using general periodical articles is fine for some research projects but is often superficial in the coverage of a given topic. Scholarly or professional publications are more comprehensive and have peer review before the articles are published. Unlike the articles in periodicals, the peer review, at the very least, helps ensure that the article meets "professional standards." Keep in mind, however, that the professional standards assurances ensure that the related studies or research were conducted following the scientific method, but does not ensure or guarantee that the respective findings, conclusions, or interpretations are accurate.

- Give yourself plenty of extra time for the library research stage. This extra time may be critical for locating a potentially important article that you have identified through a computer search of the library indexes. The article may be in a journal that is missing (unauthorized removal from the library — theft) or has been sent for binding. Most libraries send journals and serials to the bindery when they have enough to bind one full volume. The bound volumes are then placed in the stacks with the other dated journals. A good rule to follow is that if the article is six months to one year old, there is a possibility that the respective journal has been sent to the bindery. In such cases you may have to obtain the article through interlibrary loan (the library simply obtains a copy of the article from another library), which generally takes several days, or purchase the article at a higher cost from a document fulfillment center. The missing journals problem will eventually disappear as more libraries are switching to the system of having publications on compact disks, which may be full image or at least full text. The full-text articles are fine if you do not need the

Figure 2.D Sample Subject Search Screen From the Academic ASAP System at Alabama Southern Community College in Monroeville, AL

```
InfoTrac EF                    Academic ASAP              Brief Citations

  Subject: heat-transfer media
───────────────────────────── 1 of 3 ─────────────────────────────
     1         Modeling of field tests.(In Situ Mixed Region Vapor
           Stripping in Low-Permeability Media, part 3) John S. Gierke,
           Congli Wang, Olivia R. West and Robert L. Siegrist.
           Environmental Science & Technology, Sept 1995 v29 n9 p2208(9).
           Press Enter ↵ for abstract.
     2         Cooling bubbles dissipate heat.(use of microchannel flow
           boiling in cooling technology) USA Today (Magazine), June 1995
           v123 n2601 p3(2). Mag. Coll.: 79K0088.
           — Text Available —
     3         The birth of the instant cool. (new refrigeration
           technique) Jonathan Beard. New Scientist, May 8, 1993 v138
           n1872 p17(1).
           — Abstract Available —


 Display    Narrow    Explore    Esc Return to subject list
 Display extended citation       F1 Help  F2 Start over  F3 Print  F4 Mark
```

Figure 2.E Sample Full-Text (Screen 2 of 3) From the Academic ASAP System at Alabama Southern Community College in Monroeville, AL

```
InfoTrac EF                    Academic ASAP              Extended Citations

  Subject: heat-transfer media
───────────────────────────── 2 of 3 ─────────────────────────────
  Source:  USA Today (Magazine), June 1995 v123 n2601 p3(2).

   Title:  Cooling bubbles dissipate heat.(use of microchannel flow
           boiling in cooling technology)

 Subjects: Cooling - Research
           Heat-transfer media - Innovations
 Features: illustration;  photograph

 Mag. Collection:  79K0088

     AN:  17067364


 Purdue University researches are hot on the trail of a cooling technology
 that dissipates enormous amounts of heat. It could be used in the design of  ↓
 Display    Narrow    Explore    Esc Brief citation display
                                  —  Previous                    + Next
 Display brief citations          F1 Help  F2 Start over  F3 Print  F4 Mark
```

graphs and charts that were originally included in the articles. As the name full-text implies, the articles are just plain text without pictures, tables, charts, or other graphics. Again, if you find a full-text article and you need the graphic, you can request a full image of the article through the interlibrary loan system or from a document fulfillment center.

Experience has shown professionals repeatedly that library-based research is still the best source of information and data. More important, today's professionals actively

Figure 2.F.1 Sample Full-Text Article (Page 1 of 2)
Printed From the Academic ASAP System at
Alabama Southern Community College in Monroeville, AL

Database: Academic ASAP
R1   heat transfer

Cooling bubbles dissipate heat. (use of microchannel flow boiling in cooling technology)

USA Today (Magazine), June 1995 v123 n2601 p3(2).

Subjects: Cooling - Research
Heat-transfer media - Innovations

Features: illustration; photograph

Mag. Collection: 79K0088

AN: 17067364

Purdue University researches are hot on the trail of a cooling technology that dissipates enormous amounts of heat. It could be used in the design of future fusion reactors, to produce more efficient medical equipment, and to make smaller, lighter electronics.

"In many devices, from fusion reactor components and rocket nozzles to X-ray machines and computer chips, there is an enormous amount of heat being produced and a very limited area for that heat to escape from," points out Issam Mudawar, professor of mechanical engineering. "If you don't have an effective cooling technology, such highly concentrated heat can produce failure."

The amount of heat a device can be subjected to before it burns out is known as critical heat flux. For instance, a nuclear reactor exceeding its critical heat flux can trigger a catastrophic meltdown. Critical heat flux also is a major design obstacle for very small and complex electronic devices, such as computer chips, that can be damaged or destroyed if they get too hot.

Mudawar's solution is to force boiling liquid at high pressure through tiny channels the size of hypodermic needles, a process called microchannel flow boiling. Flow boiling is different from other systems that use liquid coolants because two phases of the coolant flow through the system, both liquid and vapor, instead of just the liquid phase. Two-phase cooling allows much larger increases in heat dissipation than single-phase cooling.

"When you boil liquid, you create little bubbles, like heating water in a pot. In the case of microchannels, the walls of these channels are hot because they are in contact with a heat-producing device. As we pass liquid through the channels, the liquid becomes heated to boiling. Tiny bubbles form on the channel wall and are quickly flushed out with the liquid, taking the heat away with it in the process. The instant one bubble leaves, another one forms to replace it. Every bubble that forms on the wall is like a micropump. We're essentially filling the surface with little pumps, or heat exchangers, and they're working continuously to transfer the heat from the wall to the liquid."

This type of channel technology could be used in the walls of fusion reactors to protect them from their huge heat fluxes, which could be on the order of 10,000 watts of energy per square centimeter. This is the amount of heat that would be given off by 100 household light bulbs concentrated in an area about half the size of a postage stamp.

Figure 2.F.2 Sample Full-Text Article (Page 2 of 2)
Printed From the Academic ASAP System at
Alabama Southern Community College in Monroeville, AL

Other applications that could benefit from his cooling system are lasers and medical X-ray equipment, which produce very large amounts of heat and often must sit idle until they have cooled down. Increasing cooling rates a great deal will allow them to operate more efficiently and with a larger safety margin.

One of the most promising applications for microchannel cooling is in electronics. "Computer chip manufacturers are squeezing more and more components onto a single chip, as many as 10,000,000 per chip. The more miniaturization you do, the more heat is going to be produced by the chip, so the greater the need for miniature heat exchangers."

involved in research are keenly aware that they should not and cannot stop at the library — that to conduct true comprehensive research, the researcher must look to other sources such as the Internet, CompuServe, UMI, and ERIC. True, someday, not too far off, all research resources will be available in a format that gives the researcher one-stop research, a place where books, periodicals, original reports, privately held documents, public documents, journals, and serials are all available in one place. Chances are, however, that the virtual library we hear so much about will be here in just a few years, and most likely the research college and university libraries will hold all of this technology and be the responsible keepers of the millions and millions of older printed materials that will be cost-prohibitive to digitize and make available through computer.

Internet

Although there are national political leaders taking credit for initiatives related to its creation, the "information superhighway" actually celebrated its 25th birthday on September 10, 1994. The information superhighway, more commonly known as the Internet, has evolved into a sophisticated network of computer networks since its birth in 1969. The U.S. Department of Defense established the computer network in 1969, under contract from the Department of Defense Advanced Research Project Agency, and called the network ARPAnet, which primarily served national defense facilities and major universities and colleges. Today, the Internet "links 59,000 networks, 2.2 million computers and 15 million users in 92 countries"[9] (Anthes 1995, 20). Just a few years ago, computer "wizards" at universities and colleges, were the major Internet subscribers. Today, however, the Internet is quickly becoming the new frontier to be conquered by private industry. Also, use of the Internet at home is expected to grow at almost unchartable rates in the next five to ten years.

Unfortunately, although the Internet has potential, it may be more hype than fact in terms of available services for research. In fact, as many new users learn quickly, Internet is not truly a terrific source of research documents. The reason for the exclusion of compilation and availability of research documents is linked to one very critical factor — money. Before the widespread use of Internet, research document fulfillment was handled by commercial document fulfillment centers similar to UMI; this is discussed in the next section. Primarily, these documents were ordered by telephone with the full-image document being delivered by first-class mail, overnight mail, and fax machines. The full images are digitally stored documents using up huge amounts of computer storage space. Equally important, due to existing technology limitations (sending images over the Internet in a cost-effective and efficient manner), it was simply cost-prohibitive to send these full-image documents over the Internet to the public on demand. Of course large university and corporate customers could purchase the services, which could be provided through compact disk-based technologies.

---

[9] Copyright 1995 by Computerworld, Inc., Framingham, MA, 01701—reprinted from Computerworld.

More recently, we are finding that a combination of full-image and ASCII text documents are becoming available over the Internet. Such services, however, are still handled by the document fulfillment centers where accounts must be established before ordering and receiving the documents. No doubt as these centers develop more sophisticated billing and interface systems (e.g., handling credit cards and vouchers over the Internet in a secure manner), both the full-image and ASCII text research documents will become more readily available to the average consumer and corporate users. Until then, the best and quickest way to obtain research documents will be the university and college research libraries or from the premium document fulfillment centers discussed in this chapter.

The preceding statements are supported by a survey of hard-core researchers who use the Internet, along with all other research resources, to conduct comprehensive research projects. The survey was conducted through several forums on CompuServe: (1) Education Research Forum; (2) Telecommunications Forum; and (3) Engineering Automation Forum. All these forums include high concentrations of members who are professionals in the engineering, scientific, and technical fields. The comments gathered from researchers through the survey were overwhelmingly supportive that the Internet, or the Net, does have its place in the research phases. Primarily, the Internet is best suited for contact with other researchers and colleagues and to find out about the current trends or hot topics in a given field. Also, the Internet is quickly becoming a source where customers and potential customers can obtain product information and even download computer system files and drivers on demand, which can include the purchase and downloading of complete software packages. Although the Internet's value increases almost daily, most researchers agree that it is not as good as the college and university libraries or on-line document fulfillment centers to obtain well-documented and accurate information. These findings are best supported by the comments provided by two researchers:

- John Brandt of the Educational Research Forum says:
    1. As someone involved with dissertation research, I can tell you that the Internet is helpful but not perfect. As my area is educational and school psychology, my primary search source is ERIC. There is only one library in the world that has allowed the Interneters open access to their ERIC search capabilities. For me, it is as easy to walk over to our library and do the research there on the CD-ROMs. Because the commercial group that sells the CD-ROMs and the proprietary software to run them makes their living developing this stuff...they are not willing to allow anyone access to the information free of charge. There is something called Ask ERIC on CIS, AOL, and various spots on the Internet. This contains collections of previous searches that have been popular. Most of that information is too dated for the active researcher.

        Eventually our institution will be wired with the fiber-optics necessary to allow the whole campus access. Currently we have the library holdings on-line. So the long winded answer to q1 is no, one must still use the library.

2. The major advantage of the Internet for research (at least in education) is the opportunity to contact colleagues via e-mail, listservs and usenets. By doing so one can have active access with much of the "cutting edge" information (not that there's too much cutting edge stuff in education).

3. The disadvantage to Internet is minimal. I suspect that you mean using Internet and only Internet as the source. Obviously you are limited to what is out there.

4. For me, the library has the search capability and the actual journals (or microfiche) that I need (in most cases). I don't expect that we will have complete access to all journals electronically very soon. While there are new electronic journals coming on-line every day, most are still in print form.

5. If I had the money, I'd order reprints of all of the articles I need from the reproduction services rather than feeding the copy machine. I'd likely use the on-line service and Internet as the primary source.(Brandt 1995, 1)

• Ed Huntress of the Engineering Automation Forum says:

First, "primary" research usually means digging out unpublished information, conducting your own physical research, or, in some people's definition, conducting original interviews. Is that what you mean? Doing library research usually is considered "secondary."

I've written over 330 articles, in my 20 years of writing, for trade magazines; have chapters in several books on manufacturing; and currently am writing a 600-page book on metrology. Having wended my way through Internet, SMEs BBS, and CIS, over the course of roughly 10 years, I can't imagine anything but frustration in using Internet as a major resource for research. There are some people (on PRSIG, if I remember right, or maybe on JFORUM) who keep a list of experts in various fields, who are reachable via the Internet. I used the service once, a few years ago, and realized I could get twice as much info, five times as fast, by making a few phone calls. There aren't enough of the people I want to reach who are directly accessible via the Internet; you don't know when you'll get a reply; and people usually are loathe to type as much as they'd tell you over the phone.

As for digging out articles, white papers and such from databases: yes, sometimes. I've used it mostly for publicity articles, where I'm paid well enough to justify the expense. But, after spending over $6,000 on IQuest, the most frustrating thing to me is not the cost, but the fact that the thing you want is just a little too new or a little too old, or in a journal not on the electronic databases, or something that depends on the graphics that were in the original article to make important points. I use IQuest and Dialog sometimes to get a start on a bibliography, but rarely for the main part of the work. For that, research librarians can save me much grief and exasperation. If I have time, I use a lot of interlibrary loans, of books and journals alike.(Huntress 1995, 1)

The lack of comprehensive documents and fulfillment services on the Internet does not mean that the network cannot be used in the research stage of developing a technical document. On the contrary, the Internet has several critical resources that can be used at the research stage, including:

- *Electronic Mail* — Through the Internet e-mail gateways, messages and documents can be sent internationally to almost any e-mail system, which, as a minimum, includes all Internet network users, CompuServe, Prodigy, America On-line, MCI Mail, and Sprint e-mail.
- *USENET Newsgroups* — The newsgroups are forums of individuals that have similar interests and concerns on topics or a combination of topics. A powerful feature of these newsgroups is the ability to send messages and documents to mailing lists called ListServs. The ListServe facilities send messages and documents to every member listed on the ListServe mailing list.
- *Gopher* — Developed by the University of Minnesota, gopher "allows you to browse for resources using menus. When you find something you like, you can read or access it through the Gopher without having to worry about the domain names, IP addresses, changing programs, etc." (Krol 1992, 190) (see Figures 2.G and 2.H)
- *Telnet* — Used to link or connect to other computer networks anywhere on the Internet.
- *File Transfer Protocol (FTP)* — Used to connect and transfer large files from one Internet site to another (see Figure 2.I).
- *Wide Area Information Server (WAIS)* — The WAIS helps conduct searches and obtain information from Internet databases based on key-word searches.

**Figure 2.G Main Menu From the University of Illinois
at Urbana-Champaign Gopher Site**

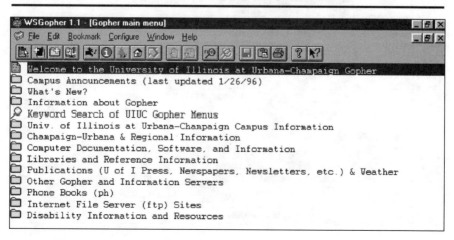

**Figure 2.H Partial Listing (One Screen) of 2570 Gopher Sites Provided
by the Gopher Services at
the University of Illinois at Urbana-Champaign**

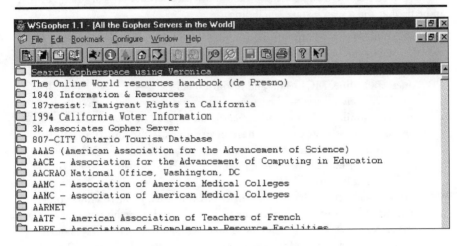

Figure 2.I FTP Screen Accessing the Microsoft FTP Site

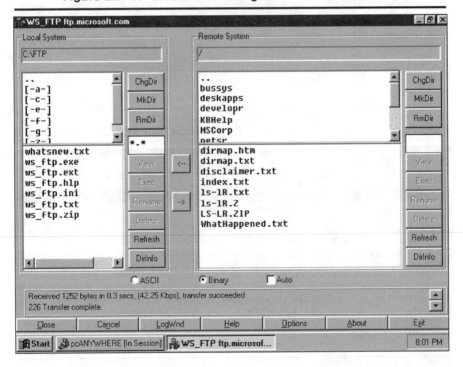

Figure 2.J Screen of the World Wide Web Through the Popular Netscape Software Program Showing the Netscape Home Page

- *World Wide Web (WWW)* — The WWW "is the newest information service to arrive on the Internet. The Web is based on a technology called hypertext. Most of the development has taken place at CERN, the European Particle Physics Laboratory; but it would be a mistake to see the Web as a tool designed by and for physicists. While physicists may have paid for its initial development, it's one of the most flexible tools — probably the most flexible tool — for prowling around the Internet" (Krol 1992, 227).

Since early 1994, the Web has grown at an extraordinary rate. Thanks to the introduction of the Web browsers such as Netscape Navigator (see Figure 2.J) and Microsoft Internet Explorer, users can access Web sites anywhere in the world by simply "clicking and pointing." This includes access to files, large documents, graphics, audio clips, and even video clips. Coupled with these browsers are the sophisticated search sites such as YAHOO (see Figure 2.K). Sites like YAHOO are comprehensive databases of Internet resources updated regularly. Users can conduct "free" searches by subjects or key-words and obtain almost instantaneous results of YAHOO findings or "hits." In turn, these findings are listed with

Figure 2.K Screen of the World Wide Web YAHOO Home Page

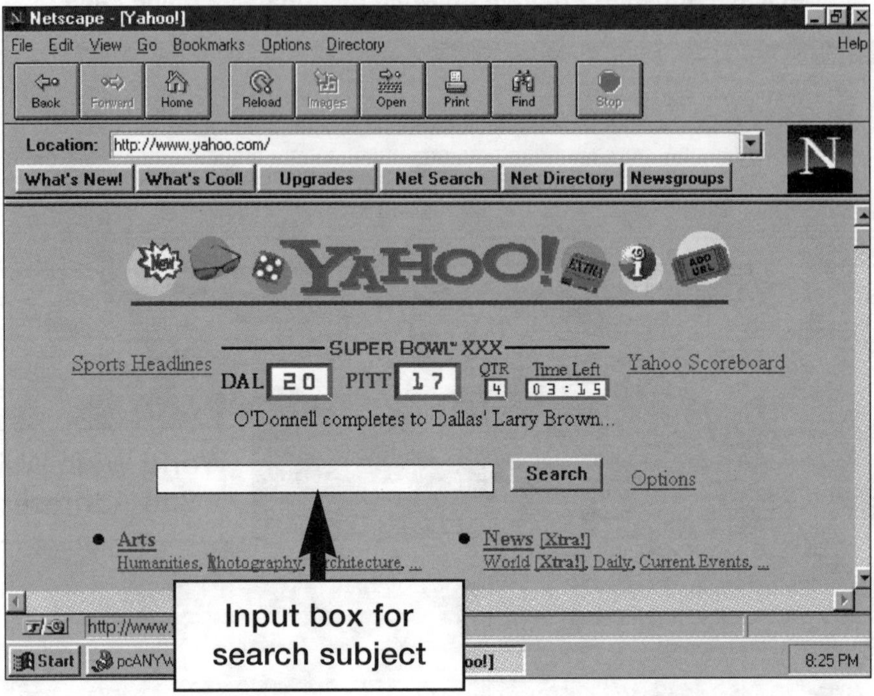

dynamic links to the respective Web site, where users simply point-and-click the desired hit and are automatically connected to the respective site.

A sample screen of a Web site is presented in Figure 2.L. A follow-up screen of the same site is provided in Figure 2.M, which provides a listing of articles that can be obtained.

Still in evolution, the Web will eventually provide users with a one-stop approach to research. Issues such as cost of providing commercial databases for use on the net, screening of professional articles, copyright protection for authors, and ensuring confidentially of credit card numbers must first be resolved before the Web becomes the "research tool" for the technical professional.

Through these resources and tools, you can undertake many research activities that significantly enhance your complete research capabilities. These activities include:

- Sending e-mail to colleagues internationally to obtain information on your specific research topic.
- Using USENET newsgroups to broadcast messages and questions to hundreds of colleagues worldwide. Questions on specific topics can be developed and sent though the respective USENET and receive as many

**Figure 2.L Screen of the North Carolina State University
Home Page for the Electric Power Research Center**

**Figure 2.M Screen of the Electric Power Research Center Publications
Available (Partial Listing) on the World Wide Web**

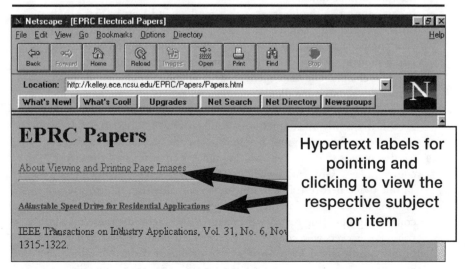

responses with added information on related projects. In some instances, it can put you in touch with international experts on the particular subject.

- Conducting broad and comprehensive searches across a majority of the 59,000 Internet networks. These searches can be conducted using Gopher, Telnet, WAIS, or WWW to locate original (generally unpublished materials that cannot be obtained through other traditional services such as document fulfillment centers) research documents. This original research document can be a tremendous source of valuable and usable information on almost any engineering, scientific, or technical topic.
- Logging in to international colleges and universities to conduct searches for books on the institution's OPAC. Generally, the larger research college and university libraries have their OPACs on their Internet link. OPAC searches can be by author, title, subject heading, or key word and can help identify books that are then obtained through interlibrary loan. Also, some of these institutions have linked their bibliographic index with the OPAC. Such configurations mean that you can conduct searches for articles held in their hard-copy stacks of journals and serials. Again, copies of these documents can be obtained for a small charge through interlibrary loan programs.

If you plan to do research through the Internet, there are several recommendations to consider that can help enhance the process and even save money and time, including:

1. Conduct your Internet searches late at night when system usage is lower than during off-peak hours, which are 12:00 a.m. to 6:00 a.m. EST. Lower system usage means that your searches and connections with other networks will be quicker.
2. If you connect to the Internet with a modem and have to make long-distance calls for the connection, faster processing speed will save you or your company a great deal of money. Of course, you should also use the fastest modem speed to help increase processing speed and save on long-distances charges. The faster speed modem will be more efficient especially during time-consuming file transfer operations.
3. When working with the USENET newsgroups be sure to review the posted, frequently asked questions (FAQ). Also, be sure that you know the interests of the newsgroup before you post any questions or comments.
4. All documents that you retrieve through the Internet are protected by copyright laws. Give credit when it is due.
5. Think twice before transferring highly confidential or proprietary information across the Internet. Although all of the Internet networks are continuing updating security procedures and methods, there may be computer hackers who could breach security. It might be best to transfer highly confidential or proprietary information from computer to computer via modems at the cost of a long-distance telephone call.

6. Do not indulge or join newsgroups that focus on hate or pornography. Predictably there is a great deal of controversy about how to allow such groups to remain without offending others and still observe all users' constitutional rights. Still, there are some groups that will take all measures to try to shut these radical groups out of the Internet — even publishing the names and company affiliations of the newsgroup subscribers. It is doubtful that your company's management would appreciate such attention.

7. When another user has responded to one of your messages or questions, always send a return note to say "thanks for the help."

8. Advertisements or related announcements should be posted only in those few newsgroups that condone or encourage such postings. Generally, a prohibitive statement related to advertising will be included in the FAQ of the respective newsgroup.

Commercially Available Computer Databases

Given today's highly technological society, there are many commercial database systems that provide access via a computer. Like anything else, the quality of these databases (i.e., timely information, full-text availability, accuracy, and scope of holdings, etc.) will depend on the access charges. Access to these databases, more properly called information services, will provide you with a wealth of information that can range from indexes, abstracts, and full-text articles from journals, magazines, newspapers, and an assorted collection of private documents and information reports. All of this information is available to you in the comfort and security of your office or home. Most of the information provided on these services dates from the mid-1980s when a majority of publishers started electronic indexing and began putting their publications in a machine-readable format suitable for use on computerized on-line services. Of course, some more exotic (and expensive) services offer full-text or abstract services of information and publications that are pre-1980.

Although there are many useful information services available, only four of the premium services are discussed below. These services include CompuServe, DIALOG, UMI, and ERIC. In total, access to the services will provide the professional with international information and data that is timely, of high quality, and will significantly enhance the development of any technical document. The only negative side of using these services will be the cost — which in certain circumstances can be quite high. On the other hand, use of these systems means sophisticated computer searches can be conducted in the comfort of your office or home and necessarily means that you will not have to travel to the nearest university research library. As a result, your time can be utilized more efficiently, instead of on traveling to and from libraries; you will not have to sift through floors of stacks; and you will not have to spend untold dollars at the copy machine duplicating articles from journals and other publications. When all expenses are totaled for the traditional library searches (i.e., your time in terms of your wages, travel expenses, and duplication expenses), the associated on-line information costs are not necessarily high. Equally important, using these on-line information services will help ensure that you obtain all of the most current available information on your research topic. Having this information or data can be extremely beneficial to your company and your career.

CompuServe

CompuServe was established in 1979 and is one of the world's largest and most comprehensive on-line services available. It is best described in an article by Van Gorder and Carter (1995, 38; Fig. 2.N):

> Membership has increased from the original 1,200 users to more than 2.5 million, with expansion of services to nearly 2,000 product areas, including news, weather and sports information, financial services, travel reservations tips and discounts, and education and reference databases.
>
> CompuServe strikes a good balance between personal and professional uses. For three consecutive years (1992 to 1994), *PC Magazine* awarded CompuServe the Editors' Choice Award in the on-line service category. In 1994, the readers of *PC World* selected CompuServe as the best on-line

Figure 2.N Initial Screen From CompuServe Information Manager

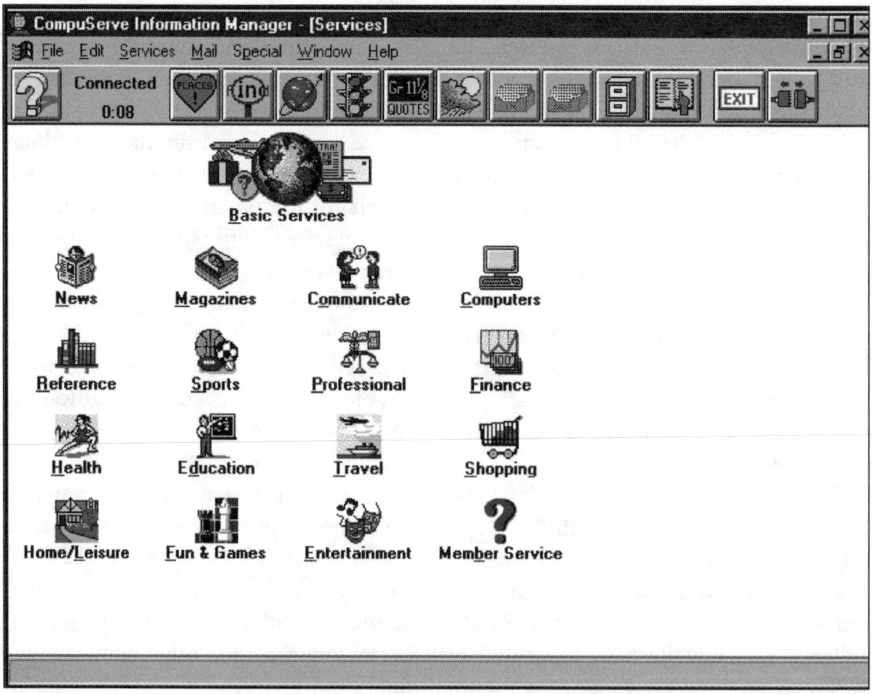

information service for the third year in a row. In addition, CompuServe was named the 1994 Readers' Choice winner in the on-line services category by *WordPerfect Magazine* and *WordPerfect for Windows Magazine*.

CompuServe's $9.95 monthly fee provides unlimited access to 100 basic services — including news, stock quotes, general reference, travel services, shopping [sic]. Members pay $4.80 an hour to access extended services, such as forums. CompuServe Mail is one of the world's largest e-mail systems. Available to nonmembers as well as to CompuServe members, it also has established ties with MCI Mail, AT&T Easylink, SprintMail, the Internet, facsimile machines, Telex and postal delivery. In addition, nearly 850 software and hardware companies provide support for CompuServe products.

Because of its comprehensive and low to moderate pricing structure (compared to other, more expensive, on-line services), CompuServe is an excellent research starting point. A sampling of its full-text and information services of particular interest to technical professionals is best described in CompuServe's *Business Reference Brochure* (CompuServe 1995, 1–5):

- *Business Database Plus*
  Keep current in your field without making trips to the library or subscribing to scores of periodicals. Business Database Plus is a service that lets you search for and retrieve full-text articles from more than 500 regional, national and international business and trade publications. Searching by word or phrase, you can locate articles that contain sales and marketing ideas, product news, industry trends and analysis.
- *Business Dateline*
  Tap into a searchable database containing the full-text articles from more than 115 regional business publications in the U.S. and Canada. You'll find a regional outlook with information on local economic conditions, retailing, real estate, people and management, financial institutions, transportation and electronics.
- *Commerce Business Daily*
  Find the latest significant federal contracts, requests for proposals, or related data in full text from the U.S. Commerce Department's publications. This database is updated daily and contains listings within the past 90 days. Information can be retrieved by subject words, sponsoring agency, ZIP code, or announcement type.
- *Corporate Affiliations*
  Investigate profiles and reference information on most large U.S. public companies and their affiliates. This database includes organizations traded on the New York and American Stock Exchanges, their affiliates, and any companies traded over-the-counter. Business information can include company name, address, phone number, business description, the names

of its executives, and the business' place in its corporate hierarchy.

- *D&B — Dun's Market Identifiers*
  Get information on over two million U.S. and 350,000 Canadian public and private companies having more than five employees or one million dollars in sales. Interested in international markets? D&B — International offers a 90-country directory on over 200,000 public, private and government-controlled companies located in Asia, Africa, Europe, the Middle East, South America, the Pacific Rim and Australia.

- *D&B — Dun's Electronic Business Directory*
  Get easy access to a directory of more than 8.5 million professionals plus public and private businesses in the United States. In addition to names, addresses, and telephone numbers, each business entry offers detailed information including a SIC code, number of employees, the populations of each entry's city and information about parent companies.

- *Legal Resource Center*
  Get the latest on major legal issues in banking and finance. Research studies covering practical and theoretical aspects of criminal justice, criminology, and law enforcement. Browse through indices to articles from over 750 law journals. Or read summaries of legislative, regulatory, judicial, and policy documents covering federal taxation. The Legal Research Center gives you access to five different databases featuring information on all aspects of law.

- *National Technical Information Service (NTIS)*
  Retrieve references or order full-text articles from government-sponsored research, development, and engineering reports. Available from 1970 to the present, most references include the title, author, corporate source, sponsor, report number, publication year, contract number and an abstract of the article. Information in the NTIS is updated every two weeks.

- *Thomas Companies and Product On-line*
  Search annually updated information on almost 150,000 U.S. and Canadian manufacturers and service providers. Also, find the latest technical information on new industrial products. Also, find the latest technical information on new industrial products introduced by U.S. (and some non-U.S.) manufacturers and sellers. Updated weekly, you'll find a wide variety of industrial products, with each database record supplying key technical data including features, attributes, and performance specifications.

- *Trademark Research Center*
  Access the premier source of all textual-numeric (nongraphic) federal trademarks active in the United States and Puerto Rico. Each entry states the U.S. class, international class, a description of the service or product, the status of the trademark, the registration data and date of first use. Additional information can include the filing date, original owner, assignor and assignee, series code and serial number.

In addition to the full-text and information services described, there are more than 2,000 special interest forums available to members and special member groups. Through these forums, individuals can access articles, files, and other critical information on topics of interest. Through these specialized forums contacts can be made with other members, which can be an equally valuable source of information — an electronic form of "networking." Of these 2,000 forums, a large percentage would be of particular interest to technical professionals. A sampling of such forums is provided in Table 2.B.

Table 2.B Sampling of Engineering, Scientific, and Technically Oriented CompuServe Forums

Forum Name	Forum Name	Forum Name
Agriculture Forum	Air Traffic Controller	AirData Forum
American Heritage Dictionary	American Public Power Assoc	Associated Press On-line
Astronomy Forum	Audio Engineering Society	Australian Associated Press
Authors Forum	Autodesk AutoCAD Forum	Automotive Information
Aviation Forum (AVSIG)	Aviation Safety Institute	Aviation Week Group
Benchmark & Standards Forum	Books in Print	British Books in Print
CADD/CAM/CAE B Vendor Forum	CADD/CAM/CAE Vendor Forum	Company Screening
Computer Animation Forum	Computer Art Forum	Computer Associates Forum
Computer Buyers' Guide	Computer Club Forum	Computer Consult. Forum
Computer Database Plus($)	Computer Library	Corel Forum
Desktop Publishing Forum	Desktop Publishing Vendor A	Desktop Publishing Vendor B
Desktop Video Forum	Electronic Books	Engineering Automation Forum
FCC Access Charge Area	FCCopy Utility Download	FORTUNE
FORTUNE Forum	FTC FREE Downloads	General Computing Forum
Government Publications	Grolier Encyclopedia	IQUEST Business InfoCenter
IQUEST Education InfoCenter	IQUEST Engineering InfoCenter	IQUEST Medical InfoCenter
IQUEST Technology InfoCenter	IQuest	IRI Software Forum
IRL Wireless Vendor Forum	ISDN Forum	Ideas & Inventions Forum
IndustryWeek Forum	IndustryWeek Interactive	IndustryWeek Management Center
Info-Please Business Almanac	Information Almanac	Information Management Forum
Information USA	Information USA Mall	Informatique France Forum
Internet Club	Internet France Forum	Internet New Users Forum
Internet Publishing Forum	Internet Resources Forum	Internet Services
Internet World Forum	MacMillan Publishing Forum	Magazine Database Plus
Masonry Forum	Nat. Computer Security Assoc	Natural Medicine Forum
Office Automation Forum	Official Airline Guide EE	PaperChase-MEDLINE
Physicians Data Query	Project & Cost Management	Project & Cost Mgt. Forum
Publications On-line	Science Museum	Science Trivia Quiz
Science/Math Ed. Forum	Sky & Telescope On-line	Society of Broadcast Eng.
Space Exploration Forum	Space Flight Download Area	Space Flight Forum
Space/Astronomy Forum	Tektronix Forum	Telecom Support Forum
Thomas Register On-line		

Source: Reprinted by permission from CompuServe, *Directory of Subjects* [on-line database (1995), Columbus, Ohio].

DIALOG

DIALOG is a comprehensive on-line service owned and operated by Knight-Ridder Information. The DIALOG computer systems are located in Palo Alto, California, and are accessible 24 hours a day. Through DIALOG, users will have access to more than "900 full-text journals, magazines, newspapers, and newsletters, besides company directories and financial data; trademark and patent information; and references to scientific and technical research" (Information, Inc. DIALOG 1991, I-1). A specific description of DIALOG provided by Knight-Ridder is as follows :

> This information has been collected, organized, and produced in machine-readable form (on magnetic tape) by many different publishers or suppliers. DIALOG contracts with the suppliers to store these collections (called databases or files) on its mainframe computers. DIALOG customers can view the information contained in over 400 databases via telecommunications networks.
>
> The database producer defines how the subject content is covered in the database and determines how the data will be formatted: as bibliographic citations, full-text articles, statistical tables, etc. The producer also provides the descriptive abstracts and develops the controlled vocabulary — key-words or descriptors or indexing terms — that enhance the accessibility of the information.
>
> Databases contain individual units of information called records. (Depending on the database, a record might be a short, bibliographic citation, a citation with a brief abstract, or the complete text of an article or report several pages long.) A single database may contain anywhere from 3,000 to 7,000,000 records. You can read the records on-line (at your terminal or personal computer) or print them out on paper. You may also use DIALOG service called DIALORDER to order actual or photocopied documents if the complete text is not available in the database. Various document delivery services accept orders via DIALORDER and deliver printed materials on request (for a fee). (Dialog 1991, 1-2 – 1-3)

DIALOG essentially has four types of databases, as follows (DIALOG 1991, 4-2–4-3):

1. *Bibliographic* — "are the most common on DIALOG. The records from these databases are bibliographic citations (title, author, source, etc.) for books, articles, magazines, etc.; they may also contain abstracts of a few hundred words that briefly describe the articles."
2. *Numeric* — "contain records which are tables of statistical data, often with text added."
3. *Directory/Dictionary* — "give factual information about companies, organizations, products, etc., often in tabular form."
4. *Full-Text* — "obtain records which have the complete text of magazine articles, newswire stories, encyclopedia entries, etc. These records may be several pages long."

DIALOG databases are grouped into 22 categories, including (DIALOG 1995, 8):

1. Business — business and industry
2. Business — business statistics
3. Business — international directories and company financial reports
4. Business — product information
5. Business — U.S. directories and company financial reports
6. DIALOG files
7. Law and government
8. Multidisciplinary — books
9. Multidisciplinary — general information
10. Multidisciplinary — references
11. News — newspaper indexes
12. News — worldwide full text
13. News — U.S. newspapers full text
14. Patents, trademarks, and copyright
15. Science — agriculture and nutrition
16. Science — chemistry
17. Science — computer technology
18. Science — energy and environment
19. Science — medicine and biosciences
20. Science — pharmaceuticals
21. Science — science, technology, and engineering
22. Social sciences and humanities

To help users with their search activities, DIALOG has developed a Windows- and Macintosh-based program called DialogLink (see Figure 2.O). For a small additional cost, users can purchase DialogLink that provides a full range of user tools from capturing images and files to production of customized accounting reports of on-line activities.

Like other on-line services, DIALOG is a for-profit business — a big for-profit business. This fact necessarily means that you will pay a premium price for its services. Again, if you follow the rule that your time is money and consider other associated research costs (e.g., time, travel costs, reproduction, etc.), you will find that even the premium costs associated with DIALOG will be cost-effective. More important, however, you will find that using premium on-line services will provide you with a scope (number of databases available in one location) and quality to help ensure that you have not overlooked vital information and have the most up-to-date information available. After all, if you do not have the most up-to-date information available on a given topic, one of your colleagues — who may not have your agenda in mind — may have it. Worse yet, a competitor who is trying to get the same contract may have this latest information.

Figure 2.0 DIALOG Search Screen Through DialogLink for Windows

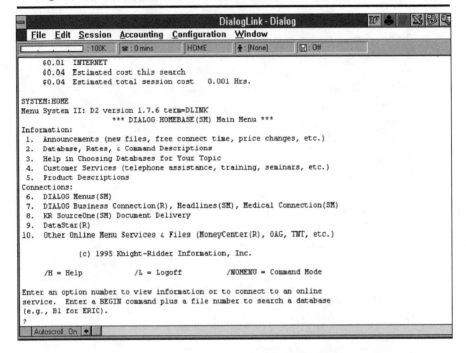

Finally, DIALOG is also linked to the Internet and a portion of the specified service is available to Internet users who have established accounts with DIALOG. More information can be obtained from DIALOG's own WWW through use of the Uniform Resource Locator (URL) Internet address: hhtp://www.dialog.com.

UMI

UMI is a one-step document source where you can obtain article reprints from more than 15,000 publications. UMI is perhaps the world's leading publication fulfillment center whose "comprehensive periodical collections" (UMI 1994, 1) is second only to the Library of Congress. All you need is the article name, publication name, and publication date and you can have the article sent in one of two ways:

1. Twenty-four-hour standard delivery — Article requested from UMI will be processed and sent to you by mail, fax, or overnight mail within 24 hours. Additionally, UMI has a service called Ariel, where full-image articles (as opposed to ASCII text) can be transmitted over the Internet. Pricing for each requested article will depend on article publication and how it is delivered. Articles sent via fax or over the Internet through Ariel will be more expensive than articles sent via first-class mail. Given this cost structure, a little planning can save you a great deal of money

while getting the articles you need without your having to travel to a library and wait in the copy-machine line.

2. Four-hour rush delivery — All of the features and services specified under the terms of the 24-hour standard delivery. Naturally, there will be a premium price associated with the rush services.

These two document-delivery options are for documents inside the UMI collection. UMI can also obtain harder to get or private papers such as technical reports or patents. The standard delivery format for such documents is two to five days. Rush jobs or same-day services are also available. Again, higher costs will be associated with these premium services.

In total, UMI has a collection of every conceivable type of document. A sampling of the types of documents available includes (UMI 1994, 1):

- Magazines and journals
- Technical reports
- SEC Files
- Government documents
- Books and monographs
- Encyclopedias
- Fliers and pamphlets
- Indexes
- AV materials
- User manuals
- Press releases
- Marketing research reports
- Software documentation
- Handbooks
- Training manuals
- Manuscripts

- Newspaper articles
- Conference proceedings and papers
- Standards and specifications
- Patents
- Theses and dissertations
- Price lists
- Directories
- Product literature and brochures
- Maps
- Translations
- Photographs
- Annual reports
- Management tapes
- Newsletters
- Advertisements

Specific subject areas serviced by UMI include (UMI 1994, 2):

- Physics
- Medicine
- Business and economics
- Engineering
- Public health and safety
- Food industries
- Management

- Chemistry
- Education
- Computers
- Industrial health and safety
- Communications
- Banking and finance
- Marketing

ERIC

An often overlooked research source is the Educational Resources Information Center (ERIC). Since its inception in 1966, ERIC has evolved into the world's largest educational information database which includes close to one million documents. Supported by the U.S. Department of Education, Office of Educational Research and Improvement, ERIC was originally established as a clearinghouse system, where each clearinghouse would collect, review, index, catalog, and distribute education-related documents, reports, and journal articles. The history and evolution of ERIC is best described by the following information sheet provided to the public:

> At the time ERIC was first discussed, the literature of education was uncontrolled. Research reports, submitted to the OE [now called the U. S. Department of Education] by their contractors and grantees, received an initial scattered distribution and then essentially disappeared. Reports from other sources generally remained equally inaccessible. ERIC was intended to correct this chaotic situation and to provide a foundation for subsequent information analysis activities and attempts to spread the use of current developments.
>
> Because of the decentralized nature of American education, the many specializations, and the existence of numerous professional organizations, ERIC's designers opted for a network of organizations rather than a single monolithic information center located in Washington. ERIC was conceived, therefore, as a network of "clearinghouses," located across the country in "host" organizations that were already strong in the subfield of education in which they would operate.
>
> Contracts with Clearinghouses [sic] originally gave them responsibility for acquiring and selecting all documents in their area and for "processing" these documents. "Processing" includes the familiar surrogation activities of cataloging, indexing, and abstracting. This scheme has worked out very well. Virtually all observers of ERIC have concluded over time that the network of Clearinghouses does a better job of identifying and obtaining the current literature of education than one single information center in Washington could ever do. With their specialized subject expertise, Clearinghouse's staff are well qualified to manage ERIC documents selection functions. Decentralization has paid off well for information analysis and user services activities. However, decentralization was not the complete answer. In order to generate products that include the output for all network components, information gathered by the Clearinghouses had to be assembled at one central place. ERIC's final design, therefore, included decentralized Clearinghouses operations integrated around a central computerized facility which services as a switching center for the network. The data recorded by each of the Clearinghouses are sent to the facility to form a central database from which publications and indexes are produced. A similar design decision was made in order to supply the public with

copies of reports added to the system. In order for ERIC to make documents available instead of just informing users that a given document existed, it was necessary to provide a document reproduction service from which any non-copyrighted document announcement could be obtained. (When permission is obtained, copyrighted materials are also reproduced.) In other words, ERIC was developed as a complete document announcement and retrieval service.

Both of these centralized services had entrepreneurial aspects. The government obviously could not afford to subsidize every user's document needs. The document reproduction effort had to become self-supporting or it would become too expensive within federal budgets. Therefore, users had to pay for reports they wanted. In the same way, dissemination of the database is not subsidized by the taxpayer; persons wanting ERIC magnetic tapes are required to meet order processing, tape, and duplication costs. The federal government limits its investments in both areas by generating a fundamental database and then permitting the private sector to market it at prices as advantageous to the public as possible.

In support of this strategy, and also because the centralized operations depended on the use of then advanced technologies (computerized photocomposition and micrographic technology), these functions were located in the commercial sector.

ERIC, therefore, emerges as a network with four levels. The first or governmental level is represented by Central ERIC (the founder, policy setter, and monitor). The second or non-profit level is made up of 16 Clearinghouses located at universities or professional societies. The third or commercial level consists of the centralized facilities and support contractors for managing databases, putting out published products, making microfiche, and reproducing documents. Fourth are the users (teachers, administrators, researchers, policy makers, counselors, parents, students, etc.) who receive the benefit of these activities. (ERIC 1992, I-5)

In a manner similar to the other premium services discussed so far, ERIC indexes and resources are accessible for a charge through several sources. First, almost all university and college libraries have access to the bibliographic ERIC indexes in either electronic form, through reference guides, or through electronic linkages such as Internet, DIALOG, CompuServe, America On-line, America Tomorrow, and GTE Educational Network Services so that articles can be identified. Except for the Internet, there is a moderate charge for searches that can be conducted by key word, subject heading, author, or date of publication. Such searches can be individual field or multiple field searches. Internet users can have free and unlimited access to the ERIC Bibliographic Indexes through Telnet at the following address: acceric@inet.ed.gov.

Once documents have been identified through a search of the ERIC bibliographic indexes and abstracts, full-text articles can be obtained through several sources. As with

Figure 2.P World Wide Web Home Page
for ERIC Reproduction Service

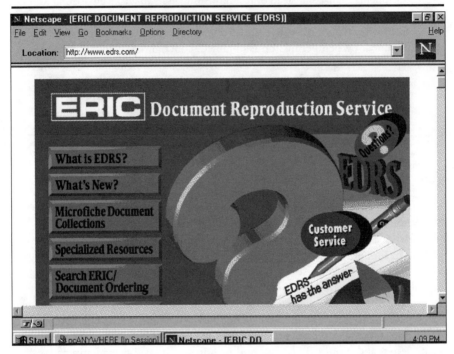

all the premium on-line services, the identified articles can be ordered with a credit card or through a preestablished account where the requested documents can be delivered via first-class mail, rush delivery via overnight delivery, or fax services. Of course, the rush orders (overnight delivery or fax delivery) will cost a considerable amount more than the standard first-class delivery. Also, documents can be ordered via telephone or fax from the ERIC Document Reproduction Services (also see Figure 20.P):

ERIC Document Reproduction
DynCorp
7420 Fullerton Road, Suite 110
Springfield, VA 22153-2852
800/443-3742
703/440-1400
WWW site: http://edrs.com

Again, though ERIC clearinghouses have an "educational twist," they are excellent sources of current published and unpublished documents on engineering, scientific, and technical subjects. At a minimum, because of the low costs of accessing the clearinghouse documents, ERIC is an excellent source of information as a first step in the research phase of your document development. Not only are these documents quality information sources on the latest trends and information on a given subject, but also an excellent resource for material on a given topic or subject.

Finally, for a small processing fee, ERIC will include most articles submitted by a willing author. Since the articles are of professional interest and quality, and documented following any acceptable industry standard including the *Chicago Manual of Style,* ERIC will process, catalog, and index your article through the appropriate clearinghouse. After all, as discussed in Chapter 1, Waddelkl indicated the importance of writing for prosperity:

> Whenever an engineer learns something new in technics, it is his bounden [*sic*] duty to put it in writing and see that it is published where it will reach the eyes of his conferes and be always available to them. It is absolutely a crime for any man to die possessed of useful knowledge in which nobody shares. (Petroski 1993, 421)

Specific procedures on getting published through ERIC may be obtained by contacting:

ERIC Processing and Reference Facility
1100 West Street
Laurel, MD 20707-3598
800/ 799-3742
301/497-4080
WWW site: http://ericfac.piccard.csc.com

Final Comments on Commercial Databases and On-Line Services

There are many other commercial databases and on-line services available in this highly technological society. The four services discussed within this chapter, however, are at the top of the list in terms of accessibility, scope of holdings, quality of services, and speed in delivery. Equally important, establishing accounts with these four services will ensure that you have full access to every conceivable type of information available internationally and that you can obtain this critical information within hours of your request. These are certainly premium services, but the associated costs will actually convert to savings in terms of your time, travel expenses, and duplication expenses. More important, having immediate access to this international information will enhance your research capabilities, which can have a tremendous positive influence on your job and career. Also, subscribing to more than one service will also provide some overlapping of services, which may have an added cost benefit as one information service on one on-line service may be cheaper on another on-line service. Thus, as in other situations, doing a little shopping and finding the best bargains in on-line services can save money.

Finally, many of these on-line services are starting to provide Internet access to their resources. Users establish an account and are given an account name and password so they can connect to the service via the Internet where the system keeps track of usage similar to when users access the system through the dial-in system. In fact, CompuServe, UMI, and DIALOG have started providing some of their services through special Internet accounts, where users can access these services through the Internet web browsers (e.g., Netscape Navigator, Microsoft Internet Explorer, etc.) used to access other WWW sites. Such practices reinforce the earlier observation that "information" is big business and will not be offered free on the Internet.

Determining Your Target Audience

Another important element of the research stage of document development is determining your target audience. As you may remember from your first-year English course, there are four main reasons why we write in a professional setting. These reasons, as related to the job of the technical professional, are as follows:

1. *To Inform* — To inform readers about important information, findings, or discoveries, which is the most basic reason for writing. Implied in this reason is to inform the readers in a manner in which they will fully understand or comprehend the findings or discoveries.

2. *To Instruct* — To transmit or teach the readers new and important information. The basic function of writing to instruct is to facilitate the transfer of knowledge so that the readers can comprehend or even perform procedures covered in the instructional material.

3. *To Persuade* — To persuade means writing to change the beliefs, attitudes, or even behavior of the readers. Because we are dealing with the emotions of the readers, this type of writing is perhaps the most difficult of the four.

4. *To Document* — To document means writing to describe how something was done. A scientist, for example, will document all steps associated with a specific experiment so it may be replicated by other colleagues for validation purposes.

Thus, the initial step associated with technical writing is to determine why you are writing in the first place. If, for example, you are assigned the task of developing a training manual on the operation and maintenance of a particular piece of equipment, then you will be writing to *"instruct."* If, on the other hand, you are assigned to develop a technical document on why your company should consider production of a new product, then you will be writing to *"persuade."*

Once the reason for writing has been established, the next step is to decide who will be your target audience. Just as a company's marketing department must determine its audience, the technical professional must determine the audience most likely to read the technical document. The target audience for a training manual on using a press on a production line, for example, would most likely be nontechnical production workers. On the other hand, the target audience for a procedure manual associated with bringing a steam generator on-line at a power plant would most likely be technicians or even engineers.

Determining your target audience is a critical step because such a determination will influence how you write. A training manual on using a press on a production line would be directed to a nontechnical audience, which means that the technical professional must avoid using highly technical words, phrases, and jargon. If such words are used, then content definitions must be provided for the convenience of the readers. In turn, if a procedure manual is being developed for bringing a steam generator on-line at a power plant, then the technical professional must use highly technical words, phrases, and jargon in a context that other technical professionals could understand. And finally, there will be instances when the technical professional must develop a technical document for mixed audiences that include both technical and nontechnical readers. Writing for mixed audiences in a manner that is not boring to the technical professional reader and does not lose the nontechnical reader is, at best, extremely difficult, and it takes a great deal of practice and experience before most technical professionals become proficient at writing for this type of audience.

Associated with the determination of who the target audience is, there are at least three other determinations or questions, including:

- Why does the reader need this information?
- What will the reader do with the information?
- What related skills must the reader have to understand the information?

Finally, the reading level of the target audience must also be determined. Determining the reading level of the nontechnical audience is especially crucial. Many technical professionals assume that because they can read at a high level, all citizens can read at the same level. Equally important, most technical professionals live in a somewhat isolated setting where they have little extended interaction with the general population and thus do not understand that most of the reading population does not read at the same level as the technical professional. This fact is best supported by Hart's observation in his article *Writing to be Read*:

> Newspapers consistently publish major stories written above the reading level of most potential newspaper buyers. Small wonder, then, that newspapers reach only half of the nation's households.
>
> Don't believe it? Just look at the facts: Only about two in five American adults have completed college. About the same proportion have graduated from high school but never went further. About one in five never finished high school. The average educational attainment is 12.7. (Hart 1993, 5)

It is also important to note that the educational attainment level (completion of a grade) is not necessarily strongly correlated with reading level. In fact, the reading level of a high school graduate is usually below the 12th-grade reading level as measured by a process called "readability."

With respect to the preceding, the remaining portions of this chapter will focus on actions that technical writers can take to help ensure they are writing technical documents for their target audience and at an understandable level for the reader.

Readability and Readability Formulas to Assist in Determining Your Target Audience

Readability is simply defined as "the ease of understanding or comprehension due to the style of writing"[10] (Roberts, et al. 1994, 119). Following this definition is the premise that all material in a technical document must be written in a style, or level, that the target audience will be able to read and comprehend. The reading levels are generally correlated to traditional grade levels where elementary and secondary reading levels are represented at specific grade levels of 1, 2, 3, 4, 5, 6, 7, 8, 9, 10, 11, and 12; and the college reading levels are represented as grade 13 to the highest possible reading level of grade 14+. More often than not, the reading level of engineers, scientists, and technicians is usually higher than the norm of the overall population. Ideally, one would expect the reading level of these professionals to be somewhere between levels 13 and 14. In the case of highly educated technical professionals holding doctorates and equally proficient in the English language, the reading level would be 14+. This does not necessarily mean that individuals who did not attend college or even who did not finish high school could not read at a level of 14+ and thus have a high comprehension or grasp of the material. Rather, it is more of a generalization reflecting the average reading level as a standardized measure of educational attainment and demonstrated reading level by a broad-based sampling of adults. That is to say that generally speaking, the higher the educational attainment level of an individual, the higher the reading level. This does not imply, however, that a high school graduate will always be able to read at level 12. Rather, it means that, generally speaking, a high school graduate will have a higher reading level than an individual who completed only the seventh grade.

A technical writer could assume that the reading level of his or her target audience is somewhat close to his or her own level, or at least close enough for the material to be understood. This does not imply that the technical writer should be oblivious to the reading level of the target audience. In fact, as discussed in the previous section, the technical writer must always keep in mind the target audience and search for the answers to three basic questions:

- Why does the reader need this information?
- What will the reader do with the information?
- What related skills must the reader have to understand the information?

If the writer can answer these three questions and develop the technical document

---

[10]Copyright ©1994, American Medical Association.

accordingly, then chances are quite high that most of the targeted technical audience will understand or comprehend the technical document material.

Determining the Reading Level of the Target Audience

Writing readable material for nontechnical and mixed audiences is much more difficult than writing for a purely technical audience. Primarily, an explicit determination must be made of the average reading level of the nontechnical audience. The ideal method of determining the reading level of a nontechnical audience is to administer a grade-level-sensitive reading assessment to a random sampling of the nontechnical audience. Such a sampling should be conducted using industry standards of sampling, and the assessment should be administered by an individual who has experience in administering and interpreting said reading-level scores. One such assessment is the *Test for Adult Basic Education (TABE),* which is produced by CTB MacMillan McGraw-Hill. The *TABE* is perhaps the basic skills assessment (reading, language, and math) most widely used in education as well as by business and industry and is designed specifically for adults. Furthermore, *TABE* comes in a computer-scorable format and, as illustrated in Figure 2.Q, the specific reading scores are presented in raw form and provide a grade-level equivalency. Administration of the reading portion of the test takes no more than 90 minutes. Most adult education programs with a local school district or a local community college use *TABE*. Also, because of their institutional missions, these schools and colleges are actually required to work with local businesses and industries, which means that they can administer the *TABE* for only a small charge to cover the replacement cost of the answer sheet.

When administering *TABE,* it is extremely important that all parties involved are sensitive to the fact that some participants will be extremely nervous about taking "a reading test." Some will be embarrassed because they know that they have low reading capability. Others will be extremely concerned about the confidentiality of the reading scores; this is especially true when you are assessing the reading scores of company employees. As a result, company officials must guarantee that assessment scores will not be reviewed by supervisors, placed in their personnel files, or used to help determine promotions or other personnel actions. Such a guarantee must be written and transmitted to the participants before they take the assessment test.

Also, if any of the employees being assessed are represented by a union, then the union leadership must be counseled on the reasons for the assessment and be presented with the written guarantee that the scores will be used only to obtain an average reading level of the group. More often than not, the union leadership will cooperate in such endeavors, provided there is an open and trusting relationship between management and the union leadership. Keep in mind the main reason for the reading assessment is to obtain the average reading level of the target audience, not to see who can or cannot read. The requirement for employees to be able to read and write at a certain level needs to be addressed during the hiring process.

Figure 2.Q TABE Report Showing Grade Equivalency Scores

See reverse side for helpful information

STUDENT MULTI-REFERENCED REPORT

ANYTOWN USD
SLADE
ANYTOWN

TABE 7&8 BASIC ED
ANYTOWN USD

Subtests	NR	SS	GE	NP	OM
READING COMP	45	662	12.9	98	100
MATH COMP	17	614	[illegible]	93	60
MATH APPS	34	630	[illegible]	97	38
LANGUAGE	43	617	[illegible]	94	50
SPELLING	18	664	[illegible]	98	100
TOTAL MATH	51	622	12.9	[illegible]	
TOTAL BATTERY	139	634	12.9	[illegible]	
READING LEVEL	A7				
MATH LEVEL	A7				
LANG/SPL LEVEL	A7				

NATIONAL PERCENTILE RANK — Below Average / Average / Above Average

1 · · 10 · · · · · 23 · · · · · 40 · · · 50 · · · 60 · · · · · 75 · · · · · 90 · · 99

Stanines 1 2 3 4 5 6 7 8 9

NR=NUMBER RIGHT SS=SCALE SCORE
GE=GRADE EQUIV NP=NATIONAL %ILE
OM=OBJECTIVES MAST

OBJECTIVE	RIGHT POSS	% CORRECT STD
READING COMP		
INTERP.GRAPHICS	10/10	100
WDS. IN CONTEXT	5/5	100
RECALL INFORMA.	5/5	100
CONSTRUCT MEAN.	13/15	87
EVAL/EX.MEANING	12/15	80
SUBTEST AVG		90
MATH COMP		
DECIMALS	5/5	100
FRACTIONS	5/5	100
INTEGERS	4/5	80
PERCENTS	2/5	40
ALGEBRAIC OPER.	1/5	20
SUBTEST AVG		68
MATH APPS		
NUMERATION	3/5	60
NUMBER THEORY	4/6	67
DATA INTERPRET.	8/10	80
PRE-ALG/ALGEBRA	3/6	50
MEASUREMENT	3/6	50
GEOMETRY	4/6	67
COMP.IN CONTEXT	4/5	80
ESTIMATION	5/6	83
SUBTEST AVG		68
LANGUAGE		
USAGE	14/15	93
SENTENCE FORMA.	9/10	90
PARAGRAPH DEVEL	10/10	100
CAPITALIZATION	2/5	40
PUNCTUATION	3/7	43
WRITING CONVEN.	5/8	63
SUBTEST AVG		78

OBJECTIVE	RIGHT POSS	% CORRECT STD
SPELLING		
VOWEL	6/7	86
CONSONANT	7/7	100
STRUCTURAL UNIT	5/6	83
SUBTEST AVG		90
TOTAL AVG		[illegible]

Grade equivalent (GE) indicates the grade placement of a student for whom a given score is typical, For Example, a GE of 12.9 means that this score is typical for a student in the ninth month of Grade 12

BATCH 960741-17 ID NUMBER 546612630
TEST DATE: 11/29/96 CODES A-J
RUN DATE: 12/09/96 PAGE 1

Published by CTB MacMillan/McGraw-Hill, Monterey, CA 93940
© Data Guide System Inc. 1984

84902
Form T-03

Conducting the reading assessment is the only accurate way to determine the reading level of the target audience. Alternative methods such as reviewing personnel records to determine reading level based on highest grade completed will not be useful in determining the actual reading level of employees.

Obviously, conducting reading-level determinations of company employees will be much easier because of accessibility than determining the same for employees of a customer or another sector of the population. The main point, however, is that a sampling of reading-level scores must be obtained to write the technical document at a level that can be comprehended by a nontechnical audience or the nontechnical portion of a mixed audience.

The average reading level scores of the nontechnical audience will then be used to help ensure that the technical document is written at the proper reading level. Just 10 years ago, writing a technical document at a specific reading level was an extremely difficult task and required the assistance of English specialists who could dissect portions of a document and conduct a complicated analysis, including counting words per sentence, counting syllables per word, and, in some cases, paragraph length. Today, thanks to the grammar checkers on the market, a reading level of a given portion of text can be determined in a few seconds by the push of a computer key. Through this analysis, a writer can determine the exact reading level of a sampling of text in the technical document. And, with experience, the technical writer will learn to modify the text to correlate with the average reading level that was determined through the *TABE*. Determination of the text reading level is conducted through a combination of three tests: (1) Flesch-Kincaid Grade Level, (2) Flesch Reading Ease, and (3) Gunning's Fog Index. All three of these tests are included in the grammar checker program called *Grammatik5™*. The description of these three tests as described in the *Grammatik5™ User's Guide* (Reference Software International 1992, 120–121) is as follows:

- **Flesch-Kincaid Grade Level** — The formula:
  0.39 × (average number of words per sentence)
  + 11.8 × (average number of syllables per word)
  Total - 15.59 = Grade Level

  A readability score of between 6th and 10th grade is considered most effective for a general audience. A higher grade level score probably means that most readers would not find the writing easy to understand. If you are writing for a special audience (e.g., a scientific or scholarly audience), a higher readability score may be appropriate.

- **Flesch Reading Ease** — The formula:
  1.015 × (average number of words per sentence)
  + .846 × (number of syllables per 100 words)
  206.835 - Total = Flesch Reading Ease Score

  The Flesch Reading Ease score is on a scale of 0 to 100. The lower the score, the more difficult the writing is to read, as shown in the following table:

Score	Reading Difficulty	Grade Level
90–100	Very easy	4th Grade
80–90	Easy	5th Grade
70–80	Fairly easy	6th Grade
60–70	Standard	7th–8th Grade
50–60	Fairly difficult	Some High School
30–50	Difficult	High School–College
0–30	Very difficult	College level and up

- **Gunning's Fog Index** — The formula:
  (average number of words per sentence)
  + (number of words of 3 syllables or more)
  Total × 0.4 = Fog Index
  The Fog Index is another measure of the approximate grade level a
  reader must have achieved to understand the document.

In addition to the three readability measures, Grammatik™ provides a correlated readability measure for each of the three scores as compared to three standard works: (1) the Gettysburg Address; (2) "The Snows of Kilimanjaro" (a Hemingway short story); and (3) a life insurance policy. These three correlations or comparisons help put the readability measures of the associated text material in perspective and are valuable when trying to write at certain grade levels.

Figures 2.R.1 and 2.R.2 show the results of the Grammatik™ readability measures for the first page of Chapter 1 of this publication. Additionally, Figures 2.R.3 and 2.R.4 show the three readability scores as compared to the three standard works. As reflected in Figures 2.R.1 and 2.R.2, the Flesch-Kincaid Grade Level measure for the text is 18, while the Flesch Reading Ease measure is 18.

There are of course, other readability programs that technical writers can use to help

**Figure 2.R.1 Grammatik™ Document Statistics Screen (Screen 1)
Showing Readability Statistics**

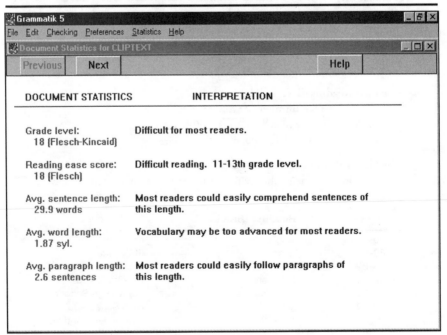

**Figure 2.R.2 Grammatik™ Document Statistics Screen (Screen 2)
Showing Readability Statistics**

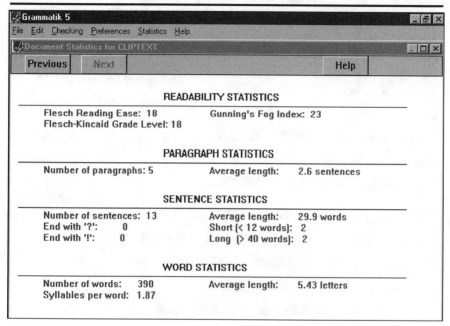

ensure that their text is written at the appropriate level. Microsoft Word 6.0 and above for Windows, for example, also has a grammar checker that provides readability statistics. In fact, the Microsoft Word program also provides readability measures by using the Flesch Reading Ease and the Flesch-Kincaid Grade Level formulas.

Using Writing Purpose, Reading Assessments, and Readability Measures to Write to the Target Audience

First and foremost, one must understand that writing is more of an art than a science, which is best described by Hart:

> Writing is an art, not a science, and you can't measure its impact via mathematics or with machines. You could write pure nonsense that would score well on a readability test. "Twas brilling and the slithy toves," would probably do just fine. Nor will readability scores tell you whether you've littered a news story with clichés and journalese, left gaping holes in the narrative or ignored vital questions.
>
> But readability tests can offer valuable guidance for thinking writers . . .
> (Hart 1993, 5)

Figure 2.R.3 Grammatik™ Document Comparison Screen (Screen 1) Showing Readability Statistics

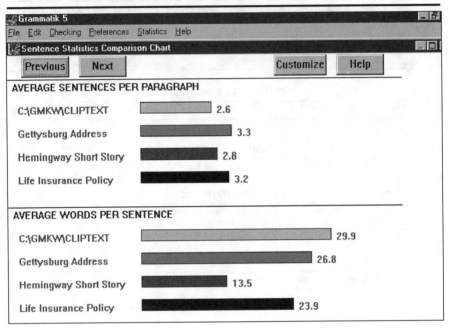

True, readability measures will not guarantee that your text will be fully comprehended by the reader. Using the readability measure coupled with average reading scores of the target audience for nontechnical and mixed audiences, and keeping the purpose for your writing in mind, will help ensure that you are at least in the "ball park." That is to say, that at least you will have an idea of where your target is and how close you are to matching the reading level of the target audience. These advantages alone make it well worth the time and effort to use these tools to help ensure that you are writing at or near the appropriate reading level of the target audience.

Along with readability measures, keep in mind that there are additional, and equally important, factors to consider in determining the readability level of a technical document. One such factor includes how many illustrative graphics are used and how. In reality, you could have a technical document section written a little above the average reading level of the target audience with properly used graphics, tables, and other illustrations to help clarify the text and obtain the same or even better results than aiming for the reading level. In fact, like anything else, there are points of diminishing returns when it comes to readability level and just how low you can bring certain technical subjects or topics. The process of hitting this point of diminishing return is called "oversimplistic and derivative prose" (Reuter 1994, 404). The fact is that to match extremely low reading levels, the writer must produce extremely short sentences with only one main clause. When combined in paragraphs, the reading

Figure 2.R.4 Grammatik™ Document Comparison Screen (Screen 2) Showing Readability Statistics

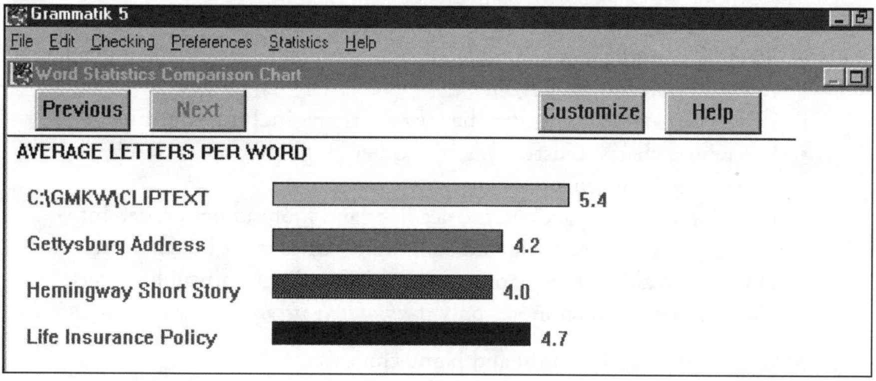

becomes so choppy that it is distracting to the reader, who loses concentration and will not, in all probability, comprehend the full meaning of the material. One way to help overcome this problem of oversimplistic and derivative prose is to "write a paragraph rather than a sentence at a time" (Reuter 1994, 405). Working on the paragraph level will help ensure that main thoughts are intact, consistent, and presented in a more digestible format than choppy sentences that are, at the very least, distracting to the reader. Again, more often than not, such problems occur only when highly technical information is being written for audiences with extremely low reading levels.

Learning to write with the target audience in mind necessarily implies that you are writing to be understood and that will take time and practice. The best way to gain such skills is to write and write some more. The more varied audiences that you must write for, the better you will become. Always keep in mind the reading level of the target audience when writing for nontechnical and mixed audiences. Using reading assessment scores and the readability measures discussed will help you stay on target.

Finally, nothing should be taken for granted. After you have developed your technical document, it is a good practice to validate its readability. The best way to do this is to select a random sample of the target audience and have them read the material. In the cases of training manuals, procedure manuals, and other documents, a simple multiple test can be developed to determine whether the readers have comprehended the material. In documents such as reports, letters, and memoranda, it is wise to have a colleague read the document and ask for constructive criticism on how the text could be improved to help enhance readability. Studies have clearly shown that such peer review helps provide better technical documents (Roberts et al. 1993, 119)[†], which means that the readers can comprehend the material.

The following are a few recommendations on writing for each of the three kinds of audiences as related to technical documents:

[†]JAMA, 272, 2:119, ©1994, American Medical Association.

1. Technical Audiences
   - Use technical jargon sparingly and only if the target audience will understand its meaning and use.

2. Nontechnical Audiences
   - Do not use technical jargon.
   - If technical terms must be used, then include definitions.
   - Use tables, charts, illustrations, and other graphics to help clarify complicated concepts.
   - Use a lot of white space and use headings and subheadings broken into logical parts. The heading and subheading titles must be descriptive. Do not use abbreviations or acronyms in headings and subheadings.
   - Use footnotes and endnotes only if *absolutely necessary.*

3. Mixed Audiences (Technical and Nontechnical)
   - Follow the recommendations for writing for non technical audiences.
   - Add a chapter or section at the back of the technical document for intense technical information that must be included in the document. Technically oriented readers can then be directed to this last section if they need more technical information

OUTLINING AND
WRITING THE
TECHNICAL DOCUMENT

Writing the first draft of a technical document is always the most difficult and, of course, is a prerequisite to completing the second draft. In terms of today's writing technologies, which include the personal computer and fast laser printers, the second draft is usually only an edited and enhanced version of the first draft. Rarely will a technical professional write a first draft and then start completely over to write the second draft. A few years back, just before the dawn of the personal computer age, young technical professionals heard stories from the more seasoned veterans of the days when two or three different versions of a technical document would be written before the final version was completed. Except for e-mail, faxes, letters, and memorandums, completely rewriting a technical document with today's technologies is virtually unheard of. In fact, if a technical professional writes and rewrites the technical document, it is probably more symptomatic of being a poor researcher, writer, or time manager.

Introduction

One of the important rules to remember when writing a technical document is to keep the readers in mind. Every step of the writing process must, along with providing information readers need to know, be conducted in a manner that relieves them of the heavy burden of reading painfully long and cluttered technical documents. In essence, the process of reading technical documents should not be painful; after all, we are not talking about physical exercise where the saying "no pain, no gain" is the rule. Thus, developing good, easy to read, and attractive documents is the result of proper planning even before the first words of the document are presented on paper or the computer screen.

To provide some guidance to the technical professional in developing technical documents, the remaining portions of this chapter will focus on providing information and recommendations that will enhance the entire writing process from conceptualizing the document in an outline form to development of the all-important first draft.

Outline

Many technical professionals equate using the outline format for developing a technical document with their experiences in first-year English composition class in college and would rather work without the outline. Unfortunately, developing a technical document without an outline is similar to taking a cross-country trip from the East Coast to the West Coast without using a road map. Certainly, you will eventually reach the West Coast, but the trip will take a lot longer than planned.

Using an outline in developing the technical document has several critical advantages:

1. Helps organize important or key points to cover in the document.
2. Provides a graphic view of the proposed flow of information and data to help ensure that they are presented in a logical format that the reader can follow easily.
3. Provides a road map–type tool that helps the writer stay with only important items and helps prevent the likelihood of drifting from the main thesis. In other words, it can help reduce rambling on about unimportant things.

Figure 3.A Microsoft Word for Windows Outline — Normal View

4. Provides a checklist to review after the first draft is completed to ensure that all key points and information have been included in the document.

Some writers may prefer to use the standard pen-and-paper outline, which is perfectly acceptable and will get the job done. Others are beginning to appreciate some of the new computer technologies that actually integrate the outlining process with the technical document. One such technology is the outlining tools provided in Microsoft Word for Windows. As illustrated in Figures 3.A (normal document view) and 3.B (outline view), the outline tool, which is actually a whole screen view within a document, provides the placement of headings and subheadings and the subordinate text (see Figure 3.C). Equally important is the fact that the outline can be collapsed so that only the headings appear to significantly shorten the length of the document. Then rearranging the document is easily achieved by simply selecting and dragging the heading to the new location in the outline, and all associated subheadings and text are also moved. This powerful rearranging feature means that large document rearrangements can be accomplished by going into the outline mode after the document has been written and selecting the headings to move. Keeping track of a few headings, as opposed to pages and pages during the editing process when sections can sometimes be moved to other locations in the document, will save a great deal of time and effort. Thus, using an integrated outline tool such as Microsoft Word can significantly enhance

Figure 3.B Microsoft Word for Windows Outline — Outline View

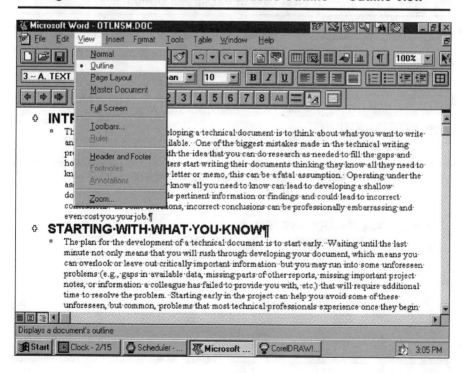

the editing process after the first and subsequent drafts are completed.

There are additional outlining tools for computers that are available, and Microsoft Word is just one of these tools. The specific tool or even how the outline is developed is not as important as using an outline. It is almost a certainty that using an outline in developing the technical document will help ensure that all necessary and critical information is included in the document, which means that information the reader needs is provided and not forgotten. Forgetting to include important, and in many cases, critical information is one of the biggest mistakes that could have been avoided by using an outline.

Staying With the Thesis

Using the outline format to develop a technical document will certainly help ensure that all important information is included in the document. Coupled with the use of an outline is the need for some sort of tool to help ensure that the writer does not wander off on a tangent, which is extremely easy to do. There are at least two

**Figure 3.C Microsoft Word for Windows Outline — Outline View
With First Line of Paragraph Option**

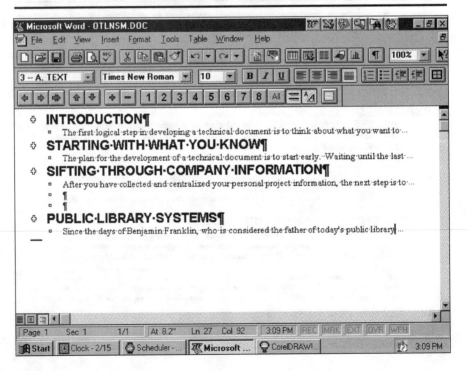

tools that can be used to help ensure that the writer stays with the thesis.

The first, which has been discussed superficially in the previous section, is to religiously use and follow the outline. The process is simple. If the outline was developed using tools similar to the outline tools provided in Microsoft Word, then the task is much easier. Simply begin writing and work your way through the document in one section; work from heading to heading, filling in the appropriate text under each heading and subheading.

Also, it is a good idea to print a copy of the entire outline before filling in the text and keep it near during the entire writing process. You can refer to this printout of the outline and other sections quickly without having to search through several pages on the computer. Following this process, you will find that as you add text and continue developing the technical document, you think of other important information that should be added under other sections and headings. By having a copy of the outline handy, you can simply pen in an addition to enter in the computer later.

Making outline additions on the printout will mean that you will not have to go to another page or section to make it, which at times may mean a little restructuring. Although making such changes is important, they can be extremely distracting when you are writing in another section of the document. Such a distraction could break your train of thought. After you have finished writing for the day, you can go back and add your penned-in notes to the computer outline. To ensure that the changes are made to the outline, it is good practice to make the changes at the end of the same day or session when you made the changes to the printout. Thus, if for some reason you lose your printout, you can always print a new copy of the outline in a collapsed format, and you will not lose important changes or additions to the computer outline.

In addition to religiously following the document outline, write down the main objectives of the technical document. That is, state in simple terms what it is you want the readers to know after reading the document. So that the objective will be easily identifiable, use a paper color (e.g., pink or blue, etc.) that you normally do not use. This way, you will be able to readily identify your document objectives sheet. Post the objectives near the area where you will be writing. If you are composing at a computer terminal, post the objectives on the far left or right side of the front of the screen. Following this procedure will ensure that you see the objectives during the writing process. Also, they will be readily available during the entire process and should be referred to every half hour or so to keep your objectives in mind as you write.

Writing in the Proper Environment

Like many other technically related tasks, there is a right place and a wrong place to do things. There is a right place to write technical documents — especially long or complicated documents. First, you must work in a quiet setting where colleagues and other workers will not be interrupting your train of thought. Some writers have the luxury of an office where they can actually close the door to keep such interruptions to a minimum. If you work in an area that has cubicles and not offices, then you must be a little creative to find a quiet spot.

An ideal quiet place would be in your home study or workroom. Today, most employers are willing to let their employees work at home under certain conditions. Writing an important report or proposal would be one of those situations. Also, most companies have laptop computers that can be taken off-site to assist in the writing process. If you are unable to work at home, then use a board room or other quiet spot to write.

Next, make certain that colleagues and other employees know to leave you alone while you are writing. If you are a supervisor, this can be a difficult task, but not impossible. Some supervisors have problems with telling their employees to leave them alone for a while. They feel the need to have and maintain an "open door" policy where employees feel free to come and talk as the need arises. However, the truth of the matter is that if the supervisor is a competent manager and runs a "smooth operation," then employees will not have to bother the supervisor with every problem immediately. Thus, the employees should be able to wait, except for emergencies, until the supervisor's writing time is over for the day.

Stopping and taking care of routine problems during the writing process is very disruptive to the train of thought. Equally important, when interacting with others, a person's heart rate increases and then the body has to come back down to a level that is conducive to the writing process. This does not imply that a writer should not attend to emergency situations that arise in the workplace, but rather should postpone common and immediately unimportant interruptions that can dramatically affect the writer's ability to concentrate and remain focused on the task. In short, it is a matter of discipline that must be practiced and followed.

Another interruption that affects a writer's ability to concentrate is the telephone. Like interacting with employees, telephone conversations can actually increase the heart rate due to the emotions of interaction. And, with telephone calls related to personnel problems, the heart rate can rise to a level that could take half an hour or more to return to a slower, more relaxed rate that is conducive to writing. Again, most people simply cannot concentrate on complex concepts and write when their heart rate is elevated. The best advice is to forward all calls to another employee when you are writing. This will help reduce interruptions that can have a negative impact on the writing process.

Finally, room temperature is also an important environmental variable that must be considered during the writing process. The ideal situation would be to work in a room temperature that is comfortable — not too hot or too cold. Working in a comfortable room temperature will help keep you more relaxed and will not be a distraction. Working in a room that is too cold or too hot is distracting, as all you can think of is the desire to make the room hotter or cooler.

How to Overcome Writer's Block

Most professional writers have faced a situation where they were unable to write a few connecting words, let alone a complete sentence. This condition is called writer's block and has been known to halt the writing process for months or longer. Hopefully such extenuating situations are rare. Even so, there will be occasions when technical professionals will experience writer's block. More often than not, such situations occur at the initial stages of writing a technical document and will be more prevalent when the technical professional is working on a major document. Career-related stress could very well be another cause of writer's block.

Career-related stress has different effects on different people and is prevalent in almost every profession. In terms of writing, the stress is usually caused by the fact that the technical professional is under extreme pressure to produce a quality technical document for the boss. Also, in some instances, the technical professional may have the unpleasant duty of writing a technical report on a project or situation where something has not gone as planned and the professional must bear part, if not all, of the responsibility.

In other situations, technical professionals may be under a great deal of personal stress involving financial problems, divorce, grief, or other personal tragedies that could also induce writer's block. Whatever the cause, every technical professional will be certain what writer's block is when it hits. One sign of writer's block is sitting at the computer or desk and not being able to write anything beyond the title of the technical document. Another sign is that of putting words to paper or the computer and realizing that they just are not right or do not say anything important. Coupled with these signs is the inability to concentrate on the task at hand, which then contributes to an even stronger case of writer's block.

The most important thing to remember is that writer's block is not, by itself, a permanent condition and usually can be cured with some simple home remedies. Also, almost every technical professional has had, at one time or another, a good case of writer's block, and it is not a sign that your career is going to end abruptly, but that conditions simply are not right for you to be creative at that particular moment.

When writer's block occurs, there are several things that can be done to help the condition pass so you can move on to the task of writing. First, try all of the

recommendations in the section on "Writing in the Proper Environment." Again, these are the known conditions that help put you in a more relaxed frame of mind so that you can begin to write.

Next, try and write something just to get your mind and body in the writing mode. Some writers go through writing exercises, similar to runners doing stretching exercises before they run. The main point is to start writing and get something on paper — even if you know what you are writing is not and could not be the final product. What you are writing at this point is not the final copy and is not carved in stone. Continue writing and more often than not the thoughts will start flowing. For some writers it may only take a few minutes before they overcome writer's block, while for others it could take as long as an hour. Once the writing flows smoothly, you can go back and refine your initial work.

If the initial writing does not work then put the paper or computer aside for a few minutes and talk to a colleague about your topic. This may help stimulate your thought processes and put you in a proper frame of mind to start writing again. When you go back to the desk or computer, write something — it is extremely important to get something in writing.

Another approach is to take a tape recorder and dictate your thoughts. Speak freely as if you were talking with a friend or family member and visualize that you are in fact "the" expert on your topic. After about 15 minutes play the tape back and transcribe what you dictated.

Finally, if all else has failed and you have not been able to write, there may be something else hindering your pursuit of writing. Specifically, there will be rare instances when you simply have not done enough research or do not know enough about the topic at hand to write about it confidently. If you suspect this may be the case, then you must obviously go back and conduct some additional research to get better acquainted with the topic. Doing additional research, however, will take time and may mean that you will not get the technical document completed on schedule. In such situations, you should first seek the counsel of the supervisor who originally directed you to write the document. Such situations would be rare and should happen only once during one's career. Generally, after a few days of additional research, the technical professional can continue with the writing process.

These are just a few recommendations on how to overcome writer's block, which affects almost every technical professional at one time or another. Of all the recommendations provided, however, the best and most effective remedy is to sit and write — put something on paper to get the process moving.

Writing the First Draft

Writing the first draft of a technical document is always the most difficult and, of course, is a prerequisite to completing the second draft. In terms of today's writing technologies, which include the personal computer and fast laser printers, the second draft is usually only an edited and enhanced version of the first. Rarely will a technical professional write a first draft and then start completely over to write the second draft. A few years ago, just before the dawn of the personal computer age, young technical professionals heard stories from the more seasoned veterans of the days when two or three different versions of a technical document would be written before the final version was completed. Except for e-mail, faxes, letters, and memorandums, completely rewriting a technical document with today's technologies is virtually unheard of. In fact, if a technical professional writes and rewrites the technical document, it is probably more symptomatic of being a poor researcher, writer, or time manager.

Assuming that all other recommendations provided in the book have been followed or at least considered, the first and most important step in completing the first draft is to start writing it. Again, the important thing is to start writing, follow your outline, and stay with the main thesis. Also, once you have the document off to a good start continue to write and write. When you reach a point where you need additional information, try to work around the section without stopping to gather the missing information. Stopping and conducting additional research once you have started writing the document can be distracting and can interrupt the work flow. In such cases, put a marker in the section where the missing information will eventually be placed. An ideal format is to place a notation in brackets and in bold, capital letters, similar to the following illustration:

[NOTE: THIS SECTION WILL BE COMPLETED AFTER ADDITIONAL INFORMATION IS OBTAINED]

Using the foregoing brackets-and-bold-capital-letters format will flag the missing information during the editing process and remind the writer that the section is incomplete. Without such a reminder it is very easy for the writer to forget. Of course, if the missing information is critical to completing other sections or portions of the document, then you must stop and get it in order to continue the writing process.

Once you have finished writing the first draft, conduct a spell check of the entire document. If you are using a large number of technical terms, abbreviations, acronyms, or proper names, the first run of the spell check may take time because some words may not be in the customized dictionary (see Chapter 5 for more information on spell checkers).

Once the spell check has been completed, then conduct a grammar check on the document (see Chapter 5 for more information on grammar checkers). Some writers will skip the spell check because the grammar checker also checks spelling . Checking spelling first may take a little extra time — perhaps as much as 30 minutes for a 100-page document — but it allows you to focus on the spelling of words in the context of the sentences. Trying to run spell checks and grammar checks in the same operation can be overwhelming and lead to mistakes because there is simply too much going on simultaneously. Finally, review the readability statistics to ensure that they are within the reading level parameters of your target audience.

The next step is to review your pagination of the document, which ensures that chapters, sections, and page breaks are at the proper points. At this point do not spend a great deal of time on inserting page breaks; just conduct a quick check to ensure that major sections, and especially chapters, break at the proper points. A more refined pagination must take place after all editing activities have been completed.

Finally, print the document and conduct an initial review for major layout problems and to ensure all desired headings are in place.

Style and Common Writing Problems to Avoid

1. **Overview.** In writing, style is the manner in which you present the material to your readers. Given this broad definition, style is all-encompassing and includes every element presented in your technical document from the voice used (active versus passive) to sentence structure (wordiness versus conciseness) to the proper choice of words and expressions (standard expressions versus clichés or colloquialisms) to the document layout (proper use of balanced white space versus a cluttered layout). With this in mind, it is a fact that some writers can combine and use all writing elements better than others and thus will be able to develop a technical document with a writing style that is easy to read and understand. On the other hand, some writers will not have a strong writing style and will not be able to use all writing elements in a proper and efficient manner, which means their documents will not always be easy to read or understand. Of course, it is important to know that most technical professionals will not be master writers, but they should and can strive to master writing. Thus, writing and improving one's writing style will be a learning experience, and in time, most technical professionals can actually have an exceptional writing style.

   In addition to proper use of all the writing elements, each person's writing style has some unique characteristics. Essentially, writing style also includes other less tangible elements that are similar to all the characteristics that affect one's personality. Thus, besides the traditional writing elements (grammar, conciseness, etc.) a writing style is a form of personal expression. This means that even with the full and proper use of all writing elements, each writer's work has a unique quality that cannot be consistently duplicated by others — a writing style as unique as one's personality. Given this uniqueness, it is extremely important that the writing style of one supervisor, for example, is not forced upon other employees, which happens more often than many professionals would believe. This does not mean that a supervisor should not strive to have his or her employees write in a way that follows all accepted rules of writing (proper grammar, spelling, writing clearly, etc.); it simply means that an employee should not have to be forced to write exactly as the "boss" does. The bottom line is that since the employees are following all accepted writing rules and company style formats (layout and form) and the documents are easy to read, then their own styles should prevail.

2. Proper Choice of Words. One of the strongest elements that contributes to an effective writing style is the proper choice of words. The choice of words and associated combinations should contribute to a document that is easy to read and clear to understand — in other words, the writer should have effectively communicated the intended meaning to the reader. Also, the choice of words should contribute to sentences that are concise. The reader should be able to read the sentence aloud comfortably, which includes the ability to read the sentence aloud without passing out because of its length.

Most style manuals include chapters on style and the proper use of words. Also, many large companies and organizations have developed their own internal style manuals that usually address some problematic areas of choosing proper words for their own particular industry or area of specialization. These style manuals should be religiously followed throughout the writing and editing process.

Explicit in the proper choice of words is the desire to avoid one common pitfall of most technical professionals. Many technical professionals believe that it is their professional duty to use several words in a sentence when only one or two are necessary. Several examples of wordy phrases that can be shortened to one or two words are illustrated in Table 3.A.

Table 3.A Wordy Phrase Versus Direct Phrase or Word

Wordy Phrase	Direct Phrase
In the event of	If
In many situations	Often
Due to the fact that	Because
In reference to our last discussion	As discussed
At this point in time	Now
In order to	To
Would appreciate it if	Please
Prior to	Before
In some instances	Sometimes

One certain way to eliminate wordy phrases is to use Grammatik™ software, which searches for these phrases and offers more direct alternatives. By using Grammatik™, the technical professional will quickly learn how to avoid using wordy phrases and choose a more direct word or phrase. An example of this type of suggestion by Grammatik™ is provided in Figure 3.D.

3. Tone. There are two basic tones in writing: formal and informal. Generally, the

**Figure 3.D Grammatik™ Interactive Window With
Suggestions on Using Direct Phrases or Words**

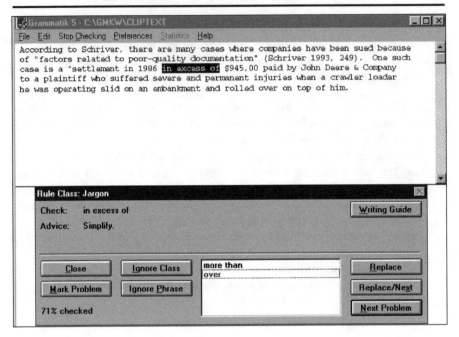

formal tone is written in the third person (neither the author nor the readers are directly addressed), while the informal tone is mostly written in the first (e.g., "I" and "me," etc.) and second person (e.g., "you," "your," and "yours," etc.). Also, tone includes other elements such as choice of words, wordiness, and document organization. Although it is hard to dispute that writing in the third person is more boring to read and, in some instances, even more difficult to read, it is proper for most technical documents. Two exceptions are for memorandums and e-mail, which could be written in an informal tone. Also, when writing for nontechnical audiences, an informal tone may be acceptable. When in doubt, however, it is best to use the formal tone for technical writing.

4. **Gender-Neutral Writing.** Since the mid-1960s, there has been much concern, and even more confusion, on how to deal with the use of gender-specific pronouns like he or she. Before the 1960s it was not only proper but standard to use the pronoun "he" even when one meant "she." Today, however, writers find themselves trying to be sensitive and politically correct and use both pronouns. Thus, in a sentence where one might have said, "The worker should take his vacation days before the end of the fiscal year," in the past decade many have been conditioned

to write, "The worker should take his or her vacation days before the end of the fiscal year." The rationale for this gender-neutral writing (eliminating sex bias) is understandable for several reasons. First, it puts both the male and female on an equal status, which is a desirable goal in all situations. Second, in today's society and workplace, women hold positions that before were only filled by men, so a writer will not be sure whether a particular situation involves a man or woman.

The only problem with using both he and she in sentences is that it becomes awkward and monotonous, especially when many technical professionals insist on using the gender-specific pronouns in an awkward and deplorable format by specifying "he and/or she." Although such usage is a generally accepted practice, it distracts the reader and relieves the writer of taking more thoughtful and creative approaches to gender-neutral writing. In essence they are saying: "There, I have written a document that is gender-neutral; that is all I need to worry about."

To complicate things even more, there has actually been a movement to revert to the days of using "he" as the correct way to indicate "he" or "she." This movement, although not really accepted by mainstream society, has managed to totally confuse many writers and readers alike. Given the "politically correct" concerns of today's society, it is doubtful that this "he" movement will influence writing in the workplace.

Since gender-neutral writing is, in all probability, here to stay, the following recommendations should be considered to reduce the awkwardness and monotony so prevalent in technical writing:

1. If it is known that the position is held by a man, use the pronoun "he" or its variation. If it is known that the position is held by a woman, use the pronoun "she" or its variation.
2. The dilemma can be eliminated by completely avoiding the use of pronouns. Although this practice is accepted in the technical disciplines, it can lead to writing boring technical documents.
3. In some instances, the pronoun "he" can be changed to "who." An example is a sentence that originally read: "If an engineer has his professional engineer license, he may transfer to another company site without loss of seniority." Change the sentence to accommodate the pronoun "who" and it will read: "An engineer who has a professional engineer's license may transfer to another company site without loss of seniority."
4. Try to avoid using the phrase "he and/or she," which is distracting.
5. Do not use the pronouns "he" or "she" in a fused format such as "he/she," which is extremely distracting to the reader. Rather, rewrite the sentence to use other pronouns such as "they" or "them," or avoid using pronouns in the sentence.

5. Trade Names Versus Common Nouns. Over the years, popular trademark names (e.g., Xerox, Kleenex, LaserJet, etc.) of everyday products have become an integral part of our language. Often these trademark names have been used in technical documents as if they were common nouns. One that is most prevalent is the misuse of the trademark name of Xerox, where people say "Make me a Xerox" when they actually own a Canon copy machine. Using trade names as common nouns is not only wrong, but also has significant legal implications for technical documents that are widely distributed. Some examples of the improper usage of trademark names and suggestions on how to present more acceptable nouns are provided in Table 3.B.

Table 3.B Use of Trade Names as Common Nouns

Trade Name	Noun
Xerox	Photocopy
LaserJet	Laser printer
Reynolds Wrap	Aluminum foil
Liquid Paper	Correction fluid
Scotch Tape	Transparent tape
Kleenex	Facial tissue

6. Numbers. Dealing with numbers in technical documents does give some technical professionals problems. Specifically, some technical professionals have trouble remembering some of the basic rules for working with numbers and associated number symbols, such as percents. Part of the problem, however, is the fact that some technical professionals get confused from information that they received in their first-year college English courses or in work situations where technical professionals were in the social sciences rather than in a technical, engineering, or scientific (physical or biological) setting.

The reason for the confusion is that some writing style manuals for the social sciences deal with numbers and associated symbols in a different manner from that of style manuals for the technical, engineering, or scientific fields. An example of such a difference is the manner in which percentages are presented in the social sciences: spelling out "35 percent" rather than showing it as "35%" as in scientific fields. Technical professionals become confused when they remember that they were taught (in first-year English) to spell out the word *percent*, and in their current field they see that the percent symbol is used instead.

There are other subtle differences between dealing with numbers and symbols in the social sciences and in the technical, engineering, and scientific (physical and biological, in particular) fields. Thus, the ideal situation is to review a style manual that is specifically written for the respective engineering, scientific, or technical discipline or field. There are, of course, several basic rules that can be summarized for dealing with numbers and associated symbols in technical writing. These rules are as follows:

- Use the percent symbol to show percents.
- Spell out numbers below 10 in most situations.
- Write numbers as figures for most situations when the number is 10 or higher.
- Spell out numbers at the beginning of a sentence; decimal numbers, however, should never begin a sentence.
- All decimal fractions less than 1.0 should be written with an initial zero (0.75, 0.24, and 0.5, etc.).
- Page and identification numbers should be written as figures.
- When using mixed numbers (i.e., whole numbers and fractions), use a hyphen to separate the whole number from the fraction (e.g., 1-1/2, 3-3/4, and 2-5/8, etc.).
- When two or more numbers appear in a series, write all of the numbers as figures (e.g., the engineer had 2 computers and 4 printers for the experiment). Spell out the smaller number, however, when two numbers appear together as a unit (e.g., the engineer had two 500 MB hard drive computers for the project, etc.).
- All numbers appearing in tables should be written as figures.
- Consult the specific discipline's style manual when using numbers in money, dates, and time.
- The decimal points of all numbers with decimals should be properly aligned as reflected in Figure 3.E.1. In most word-processing programs,

Figure 3.E.1 Table Showing all Figures Properly Aligned

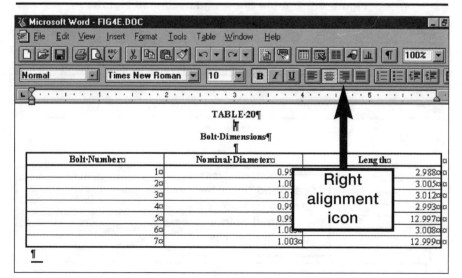

the alignment of decimals is facilitated by using the right alignment feature as shown in Figure 3.E.1. Using the right alignment feature is preferable to using spaces to try to align the decimal points; the size and locations of the numbers can make decimals a little off and create a messy, confusing document as shown in Figure 3.E.2.

7. **Mathematical Equations.** Follow the instructions in the word-processing equation program or editor as specified in Chapter 4. Also, follow the mathematical equations specifications in the style manual being used.

8. **Capitalization.** Capitalization is another area that gives technical professionals much trouble. Traditionally, the rule of many technical professionals has been "when in doubt, capitalize." Unfortunately, such a rule is not proper and, equally important, a lot of capitalization is messy and can be distracting to the reader.

To produce the most professional document, which is both pleasing in appearance and grammatically correct, there are several basic rules that should be followed in determining when to capitalize and when not to capitalize. These rules are as follows:

- Capitalize all proper nouns, including the names of persons, places, or things.
- Capitalize trademark names (e.g., Compaq, Panasonic, Ford, etc.).
- Capitalize the days of the week, months, and holidays (e.g., Monday, February, Labor Day, etc.).
- Capitalize the first word in a sentence.

Figure 3.E.2 Table Showing Misaligned Figures

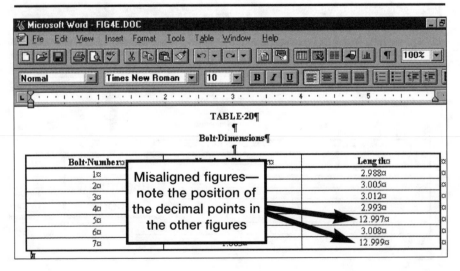

- Capitalize regions but not a specific compass point (e.g., East Coast is capitalized, while east of the city is not).
- Do not use capitals to emphasize important words or phrases (e.g., "this is IMPORTANT" would be incorrect usage of capital letters; a better approach is to put the word in italics).
- Capitalize abbreviations and acronyms of shortened forms of words that would normally be capitalized (e.g., Alabama Southern College would be ASC, Cable News Network would be CNN, or the American Society of Mechanical Engineers would be ASME, etc.). The full name should be given the first time that it appears in the text of a document, with the abbreviation appearing in parentheses immediately following the words.
- Capitalize governmental entities only if they are part of a specific name (e.g., federal government would not be capitalized, while the Federal Communications Commission would be).
- Always capitalize the pronoun *I*.
- Capitalize the first letter of all words in titles except for *a, an, and, but, for, nor, or,* and *the*.

9. **Punctuation and Style.** Because punctuation is used to help the reader better understand the written word, technical professionals must have a basic understanding of the rules associated with punctuation. Granted, there are many different types of punctuation with many more rules that should be followed. And because few technical professionals rarely memorize them all, the best rule to follow when working with punctuation is when in doubt consult the appropriate style manual for specific and detailed information on the proper use of punctuation.

There are certain types of punctuation and punctuation uses that give many technical professionals difficulty, but this can be reduced or eliminated if the following basic rules of punctuation are followed:

- *Periods*
  — Periods are used at the end of sentences.
  — Periods are used after some abbreviations.

- *Commas*
  — Commas are used to clarify a list of three or more items. Also, use the conjunctions "and" or "or" after the last comma (e.g., Supervisors' training sessions will be held on the first, second, and third Monday of each month for three months).
  — Commas are necessary to clarify the meaning of a sentence or to prevent the misreading of a sentence (e.g., the sentence "Because of the critical nature of the job engineers usually have more responsibilities than other professionals" is easier understand with the addition of one comma: "Because of the critical nature of the job,

engineers usually have more responsibilities than other professionals").
— Commas are used to set off a proper name and title (e.g., John A. Johnson, Ph.D., etc.)
— Commas are used to set off parenthetical expressions (e.g., It is evident, beyond a doubt, that the new generation of safety valves must be used on the modification of all boilers).
— Commas are used to set off dates only when presented in the day-month-year format (e.g., January 1, 1997).
— Commas are used to set off the city and state if they are both in the same sentence.

- *Apostrophes*
  — An apostrophe should be used to show truncated numbers as in years (e.g., Class of 1996 could be shown as Class of '96, etc.).
  — Watch how the apostrophe is used to indicate the possessive form of people's names ending with an *s*. The best rule of thumb to follow is to use an *s'* for more than one person's name that ends in *s* and use the *'s* for only one person's name that ends in *s* (e.g., the Sandersons'; Marcus's computer).
  — Contrary to popular trends, it is best not to use an apostrophe to show plurals of numbers. Thus the best way to show plurals of numbers is to add *s* (e.g., 1960s were turbulent years, etc.).
  — An apostrophe should be used in contractions to show letters that have been omitted (e.g., can't for cannot or shouldn't for should not).

- *Quotation Marks*
  — Quotation marks should be used to enclose direct quotations (e.g., President Nixon stated, "The end of the war is at hand").
  — Care should be taken when using quotation marks with other punctuation. One problem area is the use of quotations with commas, periods, question marks, and exclamation points. In all situations involving a direct quotation, the commas, periods, question marks, and exclamation points are placed inside the quotations (e.g., The President asked, "Is this the time to move in more ground troops?"). Also, colons and semicolons should always be placed outside the quotation marks.
  — Quotation marks should be used in the titles of journal and newspaper articles.

- *Semicolons*
  — The semicolon is used to segregate elements in a series that have been separated by commas (e.g., The main budget line items include capital outlays; computer software, hardware, and supplies; and materials and supplies).

- *Dashes*
  — The dash is used to show parenthetical emphasis (e.g., Most engineers dislike writing reports — especially long and detailed reports).

- *Question Marks*
  — The question mark is used after direct questions (e.g., Jack asked the other scientists, "What was the major finding of the study?").

- *Slashes*
  — Some writers use the slash to show acceptance of one or more terms (e.g., The computer technician will accept Microsoft Word programs and/or the program developed locally). As the use of slashes is distracting to the reader, it is best to avoid using the slash altogether (The computer technician will accept using the Microsoft Word program or the locally developed program or even both programs).

- *Italics*
  — Prior to the arrival of personal computers and word processors, the general style for using italics was divided into two separate rules. The first rule was that for handwritten and typewritten works, italics were designated by underlining the respective word or words. The second rule was for printed works: the word or words would be typeset in italics *(italics)*. Since almost every technical professional now has access to a personal computer and word processor, the use of italics will prevail in documents that are either printed following industry standards or word-processed. Handwritten documents would not, under any circumstances, be accepted in technical document situations.

With the preceding in mind, there are several related rules that should be followed for using italics, including:
  — Italics may be used to show emphasis and should be used sparingly so as to avoid monotony for the reader.
  — Italics should be used in the names of ships, aircraft, and spacecraft. Following these rules, the names of satellites and space-stations should also be in italics.
  — Italics should be used for the names of journals, magazines, and other serials (e.g., *Mechanical Engineering, ComputerWorld,* etc.).
  — Italics should be used for the titles of books (e.g., *To Kill a Mockingbird, Breakfast at Tiffany's,* etc.).
  — Italics should be used for the titles of formal reports and proposals (e.g., *Telecommunications Technologies for Distance Learning, A Proposal on the Hazard Mitigation Grant Program,* etc.).

- *Brackets*
  — Brackets are used to include information or comments to help clarify

quoted material. This would include the addition of the word *sic,* which means "so," "thus," or "in this manner." The word *sic* is used to denote that the directly quoted material is misspelled, incomplete, or incorrect, but that the author using the quote is aware of these errors (e.g., "The tests indicated that there was a variance [sic] in the findings," etc.). In other words, the author is simply stating awareness of the errors and not ownership.

PRODUCING TECHNICAL DOCUMENTS WITH A PERSONAL COMPUTER

"One of the important principles of 'Dress for Success' is that first impressions based on appearance strongly influence the reactions of interviewers, prospective clients and other audiences. This principle also applies to readers of technical reports. Contrary to the logical argument that content and correctness should matter most, readers respond first to appearances. Perhaps unconsciously, but inevitably, they look at a page as a whole visible picture before they read the words.

For busy people who read technical reports, reading is often a chore, not a pleasure, and they resent writers who make their job more burdensome by producing unprofessional-looking pages with unbroken 'gray walls' of print or sloppy typescripts with unconventual format.

Even worse than this resentment is the preconception that a report messy in design and layout is likely to be just as muddled in thought and expression. It gives the impression that the writer is inexperienced and less reliable than one who knows how to present material in an attractive, standard format" (Dolle 1990, 58).

Ray Dolle
"Designing First Impressions"
Journal of Environmental Health

Introduction

The production of a technical document, which includes everything from word processing to design and layout to printing, requires just as much attention as proper writing and editing. The truth of the matter is, however, that production of the technical document is usually the forgotten step. Usually, the technical professional will wait to focus on the production stage only after completion of the writing and editing stages. In most instances the writing and editing stages are not completed on schedule; the result is that the production process is severely compromised so that lost time cannot be made up. Unfortunately, the finished product will then be an unattractive and not very functional document. It may be just another technical document that looks like it was developed by an engineer, which means the document may never get the attention of the targeted reader. If the technical document is not attractive and functional, then the attention of the reader will never be captured. In turn, if the attention of the intended reader is never obtained, then the attention of the intended reader will never be maintained, which means that the document will never be read — regardless of how important or significant the material.

With the preceding in mind, the purpose of this chapter is to provide information and recommendations on producing the document on a personal computer to obtain a quality product that will capture the attention of the intended readers so that the document is read and understood. This purpose will be met by providing information on five critical elements in this chapter: (1) Word Processing, (2) Desktop Publishing, (3) Computerized Spelling and Grammar Checkers, (4) Document Layout, and (5) Computer Graphics. Information on the other critical elements of document production (e.g., documentation, duplication, quality control, distribution, etc.) are discussed in Chapter 12.

Word Processing

Today, word processing is a common process in every office and plant. Through word processing, individuals can create, edit, and print sophisticated documents in a variety of formats. More professionals are now using word processors to develop and produce their own documents in contrast to just 10 years ago when the same tasks were being conducted by secretaries for routine documents and by publishing houses for more sophisticated documents such as training manuals. As discussed in Chapter 1, for many technical documents, word processing will be used along with desktop publishing for the development and production of almost all technical documents. Also, as discussed in Chapter 1, to many technical professionals the line between word processing and desktop publishing is not as clear as it once was. To technical professionals who are proficient in using the personal computer for development and production of technical documents, however, the line between word processing and desktop publishing is still quite clear. Ideally, as more technical professionals begin to use both word-processing and desktop-publishing systems in the development of their technical documents, an understanding and appreciation of the differences between these systems will become clearer. For purposes of this publication, the differences between word-processing and desktop-publishing systems are best understood by reviewing current definitions and examples:

- *Word Processing Program* — A word-processing computer software package enables the user to create, edit, and print simple documents. Some higher end word-processing programs also enable users to integrate graphics, table of contents, indexing, and cross-referencing. Two of the most prominent word processing programs on today's market are WordPerfect for Windows and Microsoft Word for Windows.
- *Desktop Publishing* — Desktop publishing can merge text and graphics

on one page and also provide for full control of all page elements from kerning (spacing between characters or letters) to sophisticated text formatting around graphics. Additionally, desktop-publishing programs can incorporate full color and allow the printing of camera-ready color separations for commercial print jobs. The use of desktop publishing systems has significantly grown in recent years due to the falling costs of high-end personal computers and high-end laser printers.

Desktop-publishing systems use high-end laser printers that print at 1200 dots per inch (dpi) to 2400 dpi and also print in the encapsulated PostScript (EPS) mode. The use of EPS, which is a page description language, provides the user with greater control over printer operations and produces impressive high-quality text and graphics that are not achievable with other types of laser printers using a printer control language (PCL).

Although there are several computer software packages that have adopted the label of desktop-publishing, the two most common desktop publishing programs are Corel's Ventura Publisher™ and Adobe's PageMaker.

The use of word processing in today's workplace is a given. Ideally, the word processor will be the only software package used to develop five types of technical documents:

- E-mail
- Faxes
- Memorandums
- Letters
- Reports

The development of the remaining types of technical documents (i.e., reports, complicated procedure manuals, training manuals, and proposals) is best done using a combination of the word processor and desktop publishing. Use of these two systems incorporates two basic steps. The first step is word-processing of the text. Because desktop-publishing programs use a large amount of computer system resources and because of the manner in which text is processed, it is faster and more efficient to enter and edit the text in the word processor. Once the text has been entered and edited, it is then placed or imported into the desktop publishing program for page layout and integration of graphics. For users who are new to desktop publishing, these last steps will be a little awkward, but with time they will become comfortable.

In developing and word processing a document, most users rarely take full advantage of capabilities of the program. In today's Microsoft Windows environment which uses *Object Linking and Embedding (OLE),* tables, pictures, graphics, or other objects can be integrated into the word-processed document in two ways. First is embedding the object, which means that a copy object is fully integrated into the document. This copy can be fully edited without affecting the original object. An example of object embedding would be a graph from a spreadsheet program. By embedding the object in the word-processed document, the users have essentially made a copy of the

Figure 4.A Pasting a "Linked" Graphic in Microsoft Word

Dialog box showing option to paste the graphic as a "link"(paste link) or just as an embedded object (paste). The "link" option means that the object will be dynamically linked in the word processing document. As long as the link is maintained, any changes made to the graphic in the respective program will automatically be made to the graphic in the word processing program.

object. Then, by simply double-clicking on the object, the initial spreadsheet application will be launched (if it is a late version of a program that is properly filtered) and will allow the user to make any changes to the graph and place the edited graph back into the word-processed document. Only the copy of the object is changed — not the original graph from which the object was copied.

Object linking is similar to object embedding with one major exception. Instead of making a "detached copy" of the object, object linking utilizes a linked copy of the object (see Figure 4.A). In this configuration, the copy of the object is actually dynamically linked so that if any changes are made in the original object (called the source), then the corresponding change will be made in the word-processed document (called the client). As with object embedding, the object in object linking can be edited from the document by double-clicking on the object, which will launch the respective software program. The editing can be done and the associated changes will be made in both the source and the client objects.

Another important feature of today's word-processing programs is the ability to automate routine operations such as table of contents, title pages, quotes, lists, sublists, and other operations. This automation is simplified by style sheets that provide users with the capabilities of developing formats, as shown in Figures 4.B.1, 4.B.2, and for 4.B.3, a basic report document. When developing a style on a style sheet, the user can provide a name and select any type of text formatting from font size and style to

Figure 4.B.1 Customized Microsoft Word Style Sheet

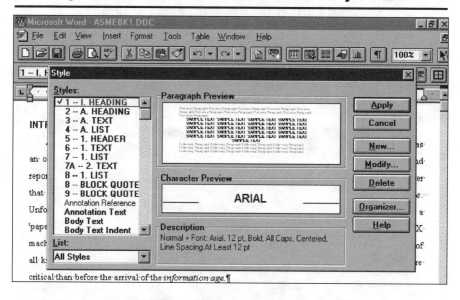

Figure 4.B.2 Customized Microsoft Word Style Sheet Showing Format for First-Level Paragraphs

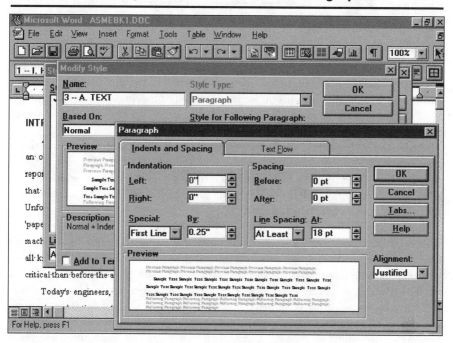

Figure 4.B.3 Customized Microsoft Word Style Sheet Showing Format for Second-Level Paragraphs

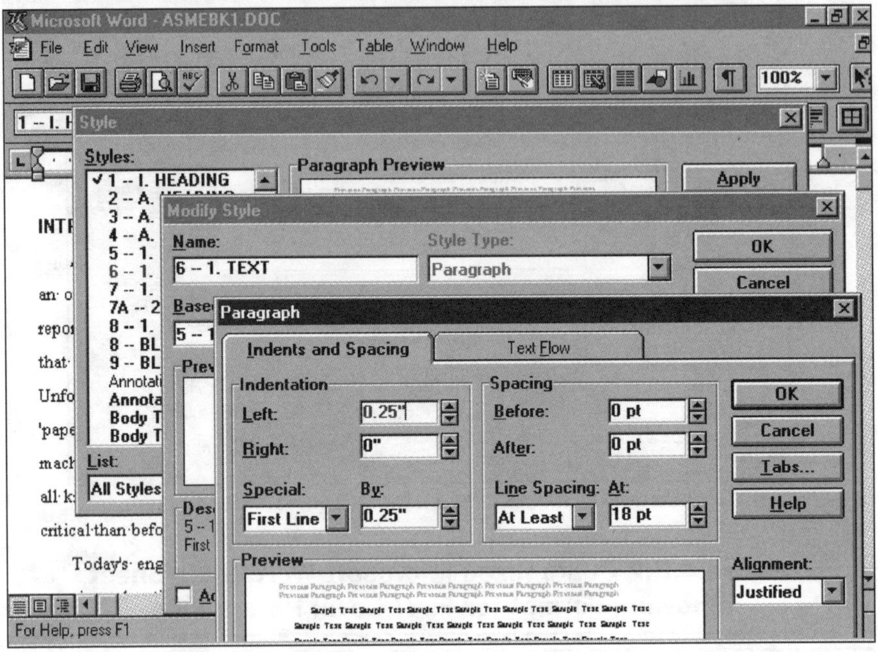

paragraph indentation to margins. The use of style sheets not only helps automate the development of the document but ensures consistency. Specifically, by using the style sheet configuration throughout the development of the document, all lists, for example, will be formatted with the same left and right indentation and line spacing. Without using the style sheets, users tend to rely on memory or try to guess what formats they used in earlier portions of the document.

Besides the style sheets, many word-processing programs include ready-made templates for routine documents. A sampling of the templates for Microsoft Word is shown in Figures 4.C.1 and 4.C.2. Many users overlook these templates, which include all the necessary styles for all formats required to develop sophisticated documents that are functional and extremely attractive.

Another important feature of today's word processors is the ability to generate a table of contents (see Figure 4.D.1). All documents of 10 pages or more should include a table of contents to help readers find critical information. The table of contents generators are semidynamic, which means that by initiating a simple command, it can be updated. As reflected in Figure 4.D.2, the creation of the table of contents is facilitated by the users inserting markers and text for each table of contents entry on the appropriate page. Ideally, the markers are placed next to or under the respective heading or subheading on the page. Customized formatting (e.g., bold, italic, lines, etc.)

Figure 4.C.1 Lisitng of 10 Templates in Microsoft Word

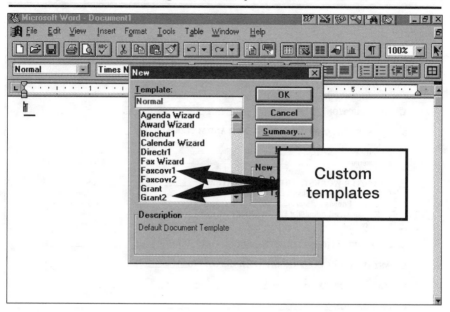

Figure 4.C.2 Listing of 10 Documents in Microsoft Word

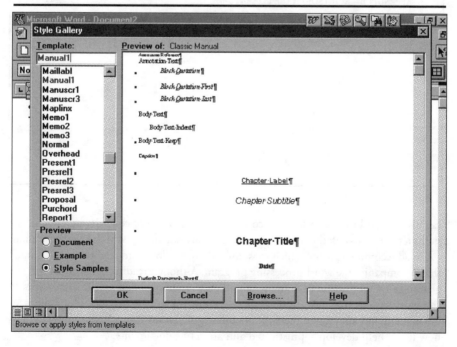

Figure 4.D.1 Table of Contents Generated With Microsoft Word

TABLE OF CONTENTS

can also be used to design table of contents levels. A table of contents in a document indicates that you took the time to develop a quality document with the reader in mind.

A good recommendation, which most users do not follow, is to read the user's manual that accompanies the word-processing program. Although reading the manual will not guarantee memorization of content, it will provide an overview of the capabilities of the word-processing program. This overview, or familiarity, will help the user understand and appreciate the full capabilities of the word-processing program and how it can help develop sophisticated and attractive documents.

Figure 4.D.2 Table of Contents Fields

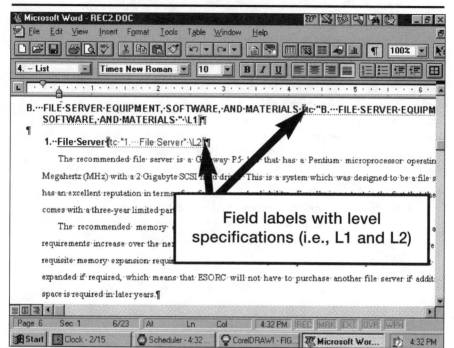

Desktop Publishing

As discussed in the previous section, desktop publishing is a sophisticated software program that provides the user with the ability to have total control of both text and graphics in a page layout, which translates to "*w*hat *y*ou *s*ee (on the screen) *is w*hat *y*ou *g*et (from the laser printer)," which computer experts call "WYSIWYG." Before the arrival of WYSIWYG, users had to be artists with a good eye for detail and spatial relationships to try and guess how the document would look in a print form instead of on the screen. New users would have to spend hundreds of hours to "get the feel" of the program before they could obtain a print copy of what they had in mind. Needless to say, WYSIWYG has made the computer less threatening and made possible the proliferation of high-end desktop-publishing programs available at low cost.

Unfortunately, the use of desktop-publishing programs still is not as prolific as it should be. Many computer users are still intimidated by these sophisticated systems and continue to produce documents through word processing that could be better produced in less time and of a higher quality through desktop-publishing systems. These systems, however, are really not that complicated. In fact, thanks to the arrival of the graphical users interface (GUI) and program features standardization brought about by Microsoft Windows, any user who has mastered a Windows-based word-processing program can master Windows-based desktop-publishing programs like Ventura Publisher and PageMaker. Because they are in a Windows environment, the touch and feel of the program menus are almost identical. In fact, at first glance the word-processing program shown in Figure 4.E.1 actually looks more intimidating than the desktop-publishing program shown in Figure 4.E.2.

Technical professionals who have never used a desktop-publishing program can start by developing small documents and gradually move up to larger and more sophisticated ones. Like anything else, the more time spent working with the desktop -publishing program, the more skilled you will become and you will soon unleash the full power of the program to develop sophisticated, functional, and quality technical documents that incorporate text and graphics.

Ideally, desktop publishing could be used to develop several types of technical documents:

- Complicated reports that incorporate both text and graphics
- Technical and training manuals that use both text and graphics in a fully integrated fashion to help illustrate the instructional material
- Procedure manuals
- Proposals

Figure 4.E.1 Microsoft Word Screen

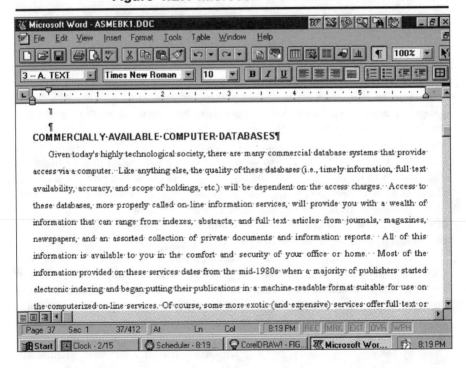

Figure 4.E.2 PageMaker™ Screen

- Newsletters
- Forms

Coupled with using a desktop-publishing program is the need to have a quality laser printer that can properly translate the program information to a high-quality printout. Today, there are many low-cost laser printers on the market at or below the $400 price level. Unfortunately, using one of these low-end printers will not produce high-resolution documents, which means the text and, more importantly, the graphics produced will be of low quality. Before purchasing a laser printer, the estimated printing volume, types of graphics that will be used, and the desired resolution should be determined. If, for example, reports or manuals will be developed that use both text and computer-aided design-type drawings of machines, parts, or architectural scenes, then a high-resolution laser printer will be required so that there is little, if any, degradation or loss of picture quality. Also, if photographs or other images will be scanned and imported into the document, then a high-resolution laser printer should be used. Ideally a 1200 dpi laser printer should be used for such applications. Additionally, such printers should be able to operate in the PostScript mode so that Encapsulated PostScript (EPS) files can be printed. If the printer does not have the ability to print in the PostScript mode, then EPS graphics and other files cannot be printed (more on PostScript files will be presented in a later section of this chapter).

Equally important, if large numbers of documents are going to be printed, then consideration should be given to purchasing a printer that prints at a higher rate (number of pages printed per minute). One of the great time wasters in the workplace is slow printing speed postscript printers that virtually lock the employee out of the computer system (particularly Windows 3.1 and 3.11) until the print job is complete. If a company does not wish to spend the additional money, which can be quite high, for a faster printer, then consideration should be given to purchasing additional printer RAM so that at a minimum all of the print file is offloaded onto the printer, which frees the computer for work.

Computerized Spell and Grammar Checkers

Almost all of today's word-processing programs have both spell and grammar checkers. And almost all of the desktop-publishing programs have spell checkers. One would think that the inclusion of these tools would mean that technical documents would no longer have spelling or grammatical errors. Unfortunately this is not so,

Figure 4.F.1 Microsoft Word Spell Checker Screen

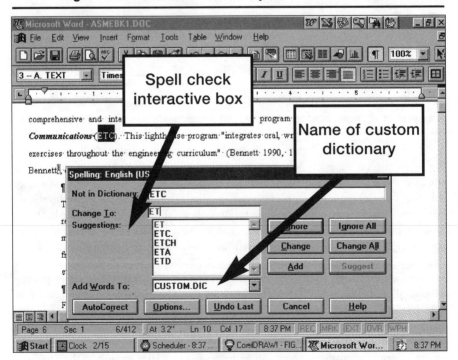

primarily because users simply do not take the time to use these extremely beneficial tools. The result is that embarrassing mistakes are made, which show the readers that the developers were careless and not concerned with quality. More important, spelling errors, which are usually typing mistakes, and grammatical errors are distracting to readers and can significantly limit the reader's comprehension of the text.

Spell checkers are usually used as the first step in editing a document. Most writers do a spell check before printing out the document. The spell checker has a comprehensive internal dictionary as a reference point and also permits users to add words to their own customized dictionaries. Also, special dictionaries (e.g., legal, medical, scientific, etc.) can be purchased and installed in most programs. The combination of internal, customized, and specialized dictionaries gives the user almost unlimited resources in ensuring that the document is virtually spelling-error free. The use of the customized dictionary is extremely important as proper names and special abbreviations can be added, so these special words will not be flagged as spelling errors. The inclusion of words in the customized dictionary can significantly enhance the speed of the spell-check process. Figure 4.F.1 shows what a spell-checker screen looks like for Microsoft Word for Windows. Figure 4.F.2 shows the contents of a customized dictionary, which is nothing more than an alphabetical listing of words

Figure 4.F.2 Microsoft Word Custom Dictionary

added during the spell-check process.

Although spell checkers are good, they are not perfect and cannot read the user's mind. An example is when a user types the word *see* in the sentence, "Joe was a simple sailor and went to see for a one-year trip." The spell checker will not flag the word or indicate that the use of "see" should actually be "sea." Another problem that occurs with spell checkers is that users are sometimes too quick to accept an incorrect word and have it included in the customized dictionary. These problems usually occur in spell checking large documents. The result is that misspelled words are added to the customized dictionary and are not flagged in later checks. One proven way to help rectify this problem is to delete the customized dictionary every few months and start over with an empty dictionary, with a new one created automatically by the word-processing program. Granted, such a practice can slow down the user as the custom dictionary is rebuilt, but such a practice can help ensure that misspelled words are not accidentally left in the document.

Coupled with the spell checkers are the grammar checkers, which help filter out the more common grammatical mistakes. Although these programs still have a way to go before they are perfect, they do help alert the users to some of the common grammatical mistakes. Equally important, the grammar checkers alert the users if

Figure 4.G Grammatik™ Screen

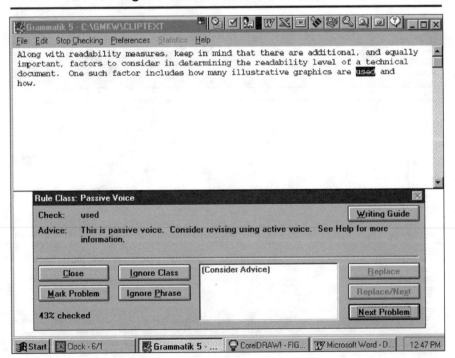

the document contains too much passive language and other boring practices such as starting every sentence with "the." Figure 4.G shows the grammar-check screen for a document in Microsoft Word for Windows. Although Microsoft Word's grammar checker is good, an external grammar checker program called Grammatik™ is excellent. Grammatik™, which is now owned by Corel WordPerfect™, has a reputation as the best grammar checker on the market. In fact, this program is so powerful that it actually helps writers improve by helping them understand the multitude of grammatical rules and practices. Finally, because this program is purchased as an external grammar checker, it will check the grammar in desktop-publishing programs as well.

Document Layout

The layout of a document is extremely important. Unfortunately, technical professionals tend not to give much thought to the document layout and actually cram as much text as possible on one page. The result is that the document looks like the writer tried to cram too much text on one page! Even worse, the technical document becomes threatening to many readers who may decide not to read it.

The best way to ensure that a technical document is read, besides writing important and useful information, is to use a proper balance of white space and text. This is one of those situations where there is no one rule that specifies what the white space to text ratio should be, except that it should be balanced and look attractive and inviting. The white space is actually used to break up the text so that reading is not tiring and there is a variety of images to capture the reader's attention. Remember one of the critical rules of technical writing: You will never keep the attention of your readers if you never get their attention in the first place. Or, as stated by White in *On Graphics: Tips for Editors*:

> Great literature needs no text breakup; people will read it in spite of its length, page after dull gray page. They know it will be worth the effort, so they approach it differently than the way they do reading matter that forms the overwhelming majority of words in print (e.g., OURS).
>
> To allure the reader into reading, we must make that prosaic prose look easy. How? By breaking it up into small pieces, for small chunks appear to ask for less commitment of time or energy on the part of the reader.
>
> A further enticement: the signals we attach to the starts of these small chunks — be they verbal (as subheads, sideheads, crossheads or whatnot) or just graphic (as initials, devices, gimmicks of some sort).
>
> Therefore, let's stop thinking of "breakup" as something negative. Instead, let's see each such break in its seductive sense — an inducement to read. Let's see each subhead for what it ought to be: that juicy, meaty, delicious gobbet of wormhood wiggling on the hook that'll help us pull that reader in (White 1993, 32).

Text can be broken up with columns, headings and subheadings, and listings of data or information—all of which are discussed in the next three subsections.

1. Columns. Columns to present information in technical documents should be used only in certain circumstances such as newsletters, instruction sheets, and certain types of training manuals. Generally, no more than two columns would be used, with at least a quarter-inch of space between the columns as reflected in Figure 4.H.1. Also, the text in columns should be fully justified, which means the text is equally aligned on the right and left margins as reflected in Figure 4.H.2.

As reflected in Figure 4.H.2, columns are very attractive in a newsletter format. In fact, producing a newsletter in a one-column format is extremely threatening to most readers and may simply not be read. Using columns in newsletters also allows the use of graphics and other objects that help break up the text.

2. Headings and Subheadings. Headings and subheadings are some of the more accepted practices to break up material. When properly used, headings and subheadings (hereafter called headings) present an organized and attractive approach to document layout. The specific heading formats, including specifications on placement (e.g., left side, centered, etc.) and style (underlined, italic, etc.) are described in all style manuals. The specific style is not as important

Figure 4.H.1 Microsoft Word Column Screen

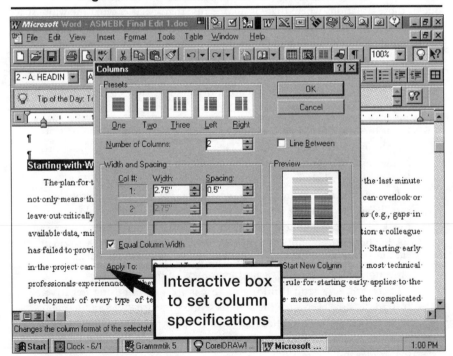

Interactive box to set column specifications

Figure 4.H.2 Document Formatted With Two Columns

STARTING WITH WHAT YOU KNOW

The plan for the development of a technical document is to start early. Waiting until the last minute not only means that you will rush through developing your document, which means you can overlook or leave out critically important information but you may run into some unforeseen problems (e.g., gaps in available data, missing parts of other reports, missing important project notes, or information a colleague has failed to provide you with, etc.) that will require additional time to resolve the problem. Starting early in the project can help you avoid some of these unforeseen, but common, problems that most technical professionals experience once they begin drafting the document. The rule for starting early applies to the development of every type of technical document from the simple memorandum to the complicated training manual.

Like most projects, one of the most critical points is starting the project. Simple logic says if you never start, then you can never finish. Traditionally, most people have trouble at two critical points: they procrastinate in starting a project and have similar problems completing it in a productive and timely manner (more on finishing will be provided in a later chapter). One of the best ways to help overcome these procrastination tendencies in beginning the project is to start early by getting organized and pulling together the information you already have. People organize their material differently, whether in file folders that are color-coded to differentiate between certain types of support

information or documentation or the more time-consuming process of using computer data bases to index all materials. Naturally, how you organize your materials is more a matter of style and will also depend on the scope of the project and the amount of information gathered. The more information you have, the more you may need computer indexing to ensure that you do not leave out any important information or data.

There are at least four types of information that you have that should be collect and reviewed. These types are as follows:

1. Letters and memorandums

2. Reports

3. Personal notes and logs

4. Personal files

SIFTING THROUGH COMPANY INFORMATION

After you have collected and centralized your personal project information, the next step is to collect company information on the project. Again, this phase should be initiated as early as possible. Usually there are at least seven types of company information that can be collected and used. These types include:

1. Source documents, including related manuals, engineering reports, drawings, and test specifications.

as the fact that the style format is consistent throughout the technical document so that readers do not become confused. Additionally, headings should be used and placed in logical locations to help highlight main thoughts or information.

3. Listing Data and Important Information. Coupled with the use of headings, the use of numbers, bullets, and other symbols should be used to help break up a lot of text if it can be logically separated. The use of numbers is appropriate when reference is made to a specific number of data elements or items.

Thus, if the text specifies "that there are four reasons why the proposal should be accepted," then the reasons should be itemized in a manner similar to the following:

1. It is cost-effective.
2. It includes all required elements.
3. It offers a simple approach to the problem.
4. It shows that the company has qualified engineers to conduct the work.

Another useful tool is the use of bullets (•) to list important information. In many instances, long text material can be broken up into smaller bulleted sections that help the reader readily identify critical information. For example, bullets can improve this information: "Complicated reports which incorporate both text and graphics, technical and training manuals that use both text and graphics in a fully integrated fashion to help illustrate the instructional material, procedure manuals, and proposals." This same information is easier to comprehend if it is presented in a bulleted format:

- Complicated reports that incorporate both text and graphics
- Technical and training manuals that use both text and graphics in a fully integrated fashion to help illustrate the instructional material
- Procedure manuals
- Proposals

The bulleted information includes spaces between the bullets and the text and includes a format called "hanging indent," where information on the second line of a specific bulleted item is in line with the beginning of the first line of text. All of today's word processors and desktop-publishing programs have the capability to produce the hanging indent style

If other information needs to be listed below a bullet or bullets, other symbols can be used. A sampling of such symbols is illustrated in Figure 4.I. Again, the main concern is to break up information in a readable and attractive format to help the reader comprehend important information and to reduce the monotony of reading large paragraphs of material.

Computer Graphics

1. **Overview.** The proper use of computer graphics can enhance the quality of any technical document. For purposes of this publication, the proper use includes selecting the correct type of graphics, placement, size, color combination, and method for reproduction of the document. More often than not, technical professionals devote all of their document development time to writing and spend little time, if any, on graphics. Sometimes, these professionals will simply cut and paste a graphic from another report. In its finished form the technical document will have the look of a cut-and-paste document. Furthermore, when the technical document is duplicated, some of the important characteristics of the

Figure 4.I Symbols That Can Be Used with Lists Created in Microsoft Word

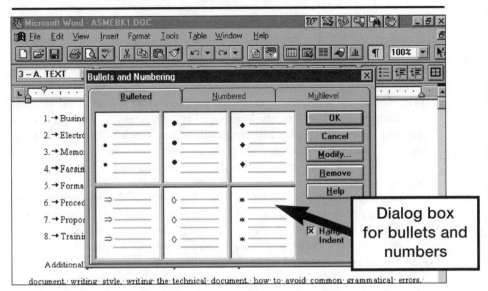

Dialog box for bullets and numbers

graphics are distorted or, worse yet, completely lost.

The information in this section provides some descriptions and recommendations on when and how to use graphics, specifically computer graphics, in technical documents. Special consideration will be given to use of gray scales with the intent that most technical documents will be reproduced in-house, using the standard copying equipment and process. Additionally, there will be some information provided on the use of color and reproduction of technical documents using color lasers and color copying machines for in-house reproduction.

2. **Graphics and Drawing Software.** Most technical professionals are now familiar with using today's word-processing programs to develop the most common technical documents, and an ever-increasing number are finding that mastery of today's sophisticated desktop-publishing programs is not a difficult task. Most technical professionals *could* master computer graphics software just as quickly as they mastered the desktop-publishing software. Again, because these software packages are Windows-based, once writers have mastered any one of the sophisticated computer programs, they can learn to handle other programs easily. Also, one of the best features about these programs is that the users do not have to be artists to develop the type of graphics that would normally be used in technical documents developed in-house.

Of course there are times when the skills of an artist will be needed, but for most technical document applications, graphics can be developed or even modified through several high-end graphics or draw-type software packages. Also, there are many vendors who sell high-quality graphics on any subject. Finally, technical drawings and photographs can be scanned and converted into graphics files for placement in the respective technical documents. Today's technical professionals have several alternatives at their disposal to develop, obtain, and place sophisticated graphics that will significantly enhance the look and feel of their technical documents. One such manner is with draw-type software packages which are completely compatible with the high-end word-processing and desktop-publishing programs.

There are several good draw-type software packages on the market that could assist the technical professional in developing quality documents. A sampling of these programs includes Microsoft Paintbrush which comes with Microsoft Windows. Although Paintbrush is not a high-end software package, it can be used to develop simple graphics or even to touch up graphics in the more standard file formats which will be discussed in the next subsection. Another program is Adobe Illustrator which is an extremely popular high-end graphics package. Also, for those situations where a large number of scanned photographs will be used and must be modified, Adobe Photoshop and CorelPHOTO-PAINT™ are excellent choices. Finally, another popular graphics package is CorelDRAW™, which is a high-end draw-type program that provides users with every tool required for development or modification of most graphics used in today's technical documents. Because CorelDRAW™ is so comprehensive, cost-effective, and easy to use, the remaining portions of this subsection will focus on it.

Currently CorelDRAW™ sells for under $400 and is a good buy because it is so comprehensive. A sampling of the program elements that are standard with CorelDRAW™ includes (Corel Corporation 1994, A–N):

- *CorelDRAW*™ — a vector-based drawing program with extensive text-handling and precision-drawing features that make it the ideal tool for virtually any design project — from logos and product packaging to technical illustrations and advertisement.
- *CorelPHOTO-PAINT*™ — a powerful paint and photo-retouching application featuring numerous image-enhancing filters for improving the quality of scanned images. Plus special effects filters that can dramatically alter the appearance of images.
- *CorelCHART*™ — a charting program for building charts and graphs of all types — from simple bar and pie graphs to three-dimensional area and pictographs. You can enter chart data from scratch into the program's Data Manager, or import files from several popular spreadsheet and database programs.
- *CorelMOVE*™ — an animation program that lets you create both simple and complex animations. Used on their own or in CorelSHOW™, animations you create in CorelMOVE™ can turn a dull presentation into a spectacular multimedia event.
- *CorelTRACE*™ — converts bitmap images (the kind that paint programs like CorelPHOTO-PAINT™ and scanners create) into vector graphics images.

- *CorelMOSAIC*™— lets you view entire subdirectories of images on screen before opening one. You can use it to store images in compressed libraries and perform batch operations such as printing and exporting groups of files.
- *CorelSHOW*™— lets you assemble printed or on-screen presentations using objects from CorelDRAW™, CorelCHART™, CorelPHOTO-PAINT™, CorelMOVE™ and other programs that support Windows Object Linking and Embedding (OLE).
- *CorelQUERY*™— is a data query facility that allows you to gather information from various data sources such as spreadsheets and databases and combine the information into tables you can sort, search and link to other applications and print.

Of the CorelDRAW™ program elements, the basic CorelDRAW™ and CorelPHOTO-PAINT™ will be used most often in developing or modifying graphics for technical documents. One of the nice features of these programs is that almost any type of computer program file can be imported into these two programs and may be modified. Such modifications can include enhancements such as adding descriptive labels and narratives. A sample screen from CorelDRAW™ is shown in Figure 4.J.

Figure 4.J CorelDRAW™ 5 Screen

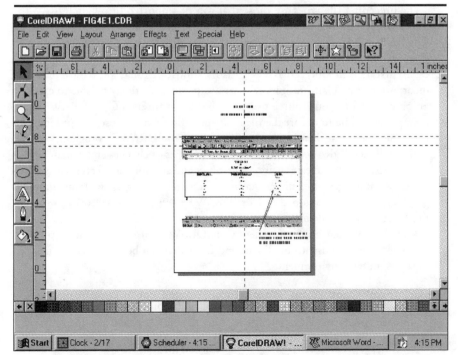

A sampling of the file types that can be created and exported from CorelDRAW™ includes:

- CorelDRAW™ (.cdr)
- CorelCHART™ (.cch)
- Windows Bitmap (.bmp, .dib, and .rle)
- Computer Graphics Metafile (.cgm)
- Kodak Photo CD Image (.pcd)
- Scitex CT Bitmap (.sct, .ct)
- TIFF Bitmap (.tif, .sep, .cpt)
- Adobe Illustrator 1.1, 88, 3.0 (.ai, .eps)
- GEM file (.gem)
- IBM PIF (.pif)
- Macintosh PICT (.pct)
- PostScript (Interpreted) (.eps)
- Ami Professional 2.0, 3.0 (.sam)
- Excel for Windows 3.0, 4.0 (.xls)
- Lotus 123 3.0 (.wk?)
- MS Word for Windows 2.x (.doc)
- MS Word 5.0, 5.5
- MS Word for Macintosh 4.0
- WordPerfect 6.0 (.wpd)
- WordPerfect 5.0 (.wpd)
- Corel Presentation Exchange (.cmx)
- CorelTRACE™ (.eps)
- CompuServe Bitmap (.gif)
- JPEG Bitmap (.jpg, .jiff, .jtf, .cmp)
- Paintbrush (.pcx)
- Targa Bitmap (.tga, .vda, .jcb, .vst)
- Windows Metafile (.wmf)
- AutoCAD DXF (.dxf)
- HPGL plotter file (.plt)
- Lotus (.pic)
- Micrografx 2.x, 3.x (.drw)
- EPS (Placeable) (.eps, .ps, .ai)
- ASCII Text (.txt)
- Lotus 123 1A, 2.0 (.wk?)
- MS Word for Windows 6.0 (.doc)
- MS Word for Windows 1.x
- MS Word for Macintosh 5.0
- Rich Text Format (.rtf)
- Word Perfect 5.1 (.wpd)
- WordPerfect Graphic (.wpg)

The preceding file types are those that can be imported into CorelDRAW™. Additionally, optional filters can be obtained for many other programs not listed. The best source to obtain the filters for other programs is to contact the Corel Corporation, which will then refer you to the appropriate source for a filter to be easily installed in CorelDRAW™. Also, as most of today's software packages incorporate cross-platform importing and exporting, you may be able to export a graphic file from your source program that has a filter in CorelDRAW™. An example would be to export the file as a ".pcx" file from your source program. In turn, the ".pcx" can then be imported into the CorelDRAW™ program.

Through CorelDRAW™, users can make all types of changes and enhancements to graphics, as well as create their own graphics and add special effects such as three-dimensional images, stretching, rotating, and pattern filling. As with the imported graphics, graphics developed in CorelDRAW™ can then be saved and exported to any of the listed file formats.

Besides all of the features specified, CorelDRAW™ also comes with a compact disk containing thousands of clip art and fonts that can be downloaded and used in the Windows environment.

3. **Types of Graphics Files.** As illustrated in the previous subsection, there are many different types of graphic file formats that can be used. For the most part, the file types are specifically related to a particular software package. File formats

of ".dxf," for example, are files created in the popular computer-aided design program AutoCAD. Another example is ".pic" files, which are created in Lotus 1-2-3. These program-specific files can be imported into CorelDRAW™, word-processing, and desktop-publishing programs with little effort. When possible, however, it is best to use one of three of the more common graphic file types: (1) Paint-Type Graphics (PCX), (2) Tag Image File Format (TIFF), and (3) Encapsulated PostScript (EPS). The main advantage of using these common graphic file formats is that generally these formats can be readily imported into any word-processing or desktop-publishing program, which provides users with greater flexibility. These same common formats can be readily imported into several presentation-type programs such as Aldus Persuasion and Microsoft PowerPoint. Through programs like Persuasion and PowerPoint, slide presentations can be developed to accompany oral presentations of technical documents. These slides can be made into high-resolution, 35-mm slides for projection, color overhead transparencies, or even converted from a computer VGA signal to a composite signal for presentation on a color television.

Definitions of the three common graphic format files as defined by Bryan Pfaffenberger in *Que's Computer User's Dictionary* (1992) are as follows:

1. *Paint File Format (PCX)* — A bit mapped graphics file format found in programs such as MacPaint and PC Paintbrush.

   The standard paint file format in the Macintosh environment is the 72-dots-per-inch format originally used by MacPaint, which is linked to the Mac's bit-mapped screen display. In Windows computers, the Windows bit map file format (BMP) is increasingly common. Many programs also recognize the PC Paintbrush file format (PCX). (p. 446)

2. *Tag Image File Format (TIFF)* — A bit-mapped graphics format for scanned images…TIFF simulates gray-scale shading. (p. 591)

3. *Encapsulated PostScript (EPS)* — A high-resolution graphic image stored in PostScript page description language. You use this language to write instructions for storing the graphic image.

   The EPS standard enables the device-independent transfer of high-resolution graphic images between applications. EPS graphics are of outstanding quality and can contain subtle graduations in shading, high-resolution text with special effects, and graceful curves generated by mathematical equations . . . EPS images can be sized without sacrificing image quality.

   The major drawback of EPS graphics is that a PostScript-compatible laser printer is required to print them — and with most application programs, the image is not visible on-screen unless a PICT- or TIFF-format screen image has been attached to the EPS file. (pp. 220–221)

As specified in the definitions, the highest–quality graphic file format will be the encapsulated PostScript (EPS) file format. The EPS format provides high-quality and high-resolution graphics which, when imported, produce crisp, clear graphics. The scaling of an EPS is extremely important. In some instances imported graphic files are sometimes too large to properly place the graphic in the technical document. Thanks to the scaleability of the EPS, however, proportional scaling can be accomplished without losing any resolution or other quality elements. The proportional scaling means that both the height and width of the graphic can be simultaneously sized so that the height-to-width proportions remain constant and the image is not distorted. Also, as specified in the EPS definition, EPS file images cannot be seen in some application programs, such as word-processing and desktop-publishing programs. If the EPS file is created or enhanced in CorelDRAW™, however, the default export EPS file format attaches an image header so that the image can be viewed in the application program. Figures 4.K.1 and 4.K.2 show an EPS file with and without the image header. In Figure 4.K.2, the image is visible, while in Figure 4.K.1, which was not exported with an image header, only the file name, creation date, and program creation name are shown.

Finally, when developing and using graphics in technical documents, special attention needs to be paid to how the document will be reproduced. Most internal

Figure 4.K.1 Encapsulated PostScript File Without Header

Figure 4.K.2 Encapsulated PostScript File With Header

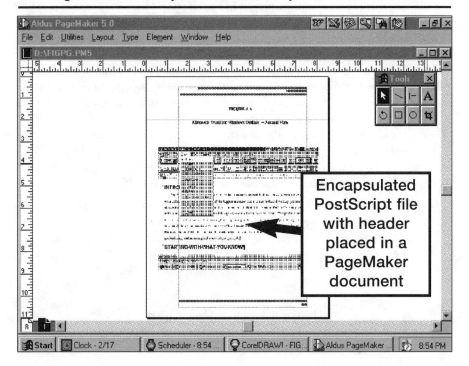

documents generated by technical professionals will be duplicated by a standard copy machine. Users must remember that due to copy machine limitations, many graphics will not be duplicated at the high-quality levels as obtained from a copy directly off the laser printer. With a little planning and testing these shortcomings can usually be overcome by adjusting the gray scales or even colors on the graphic.

The best rule to follow is do not wait until the last minute to try duplicating the graphics that will be used in the technical document. All elements of the copied graphic that have lost resolution during the copying process can then be enhanced through CorelDRAW™ or on another software package of choice. In time, experience will be gained on how to develop graphics that can be copied on a standard copy machine with little or no loss of graphics quality.

In the case of using a color laser printer and a color copier, the same basic rule follows: do not wait until the last minute to check graphics reproduction quality. Also, in working with color copiers, special attention should be given to using extremely thin color lines as they do not always print clearly and give the document a blurry effect.

4. Selection of Graphics to Illustrate Data — Graphs and Charts. A common mistake that technical professionals make in developing technical documents is that they will spend paragraphs and even pages to describe information and data that could be more easily explained with graphs and tables and less text. In short, they fail to follow the old saying that "one picture is worth a thousand words."

When used properly, graphs and tables can help highlight the important information and help support the thesis of the related text; graphs and text complement one another. Putting all graphs and tables and no explanation is just as distracting and confusing to the reader as not using any graphs and tables. In short, there must be a balance of the proper amount of explanation that should accompany any graph or table. Although there is no magic formula, common sense should be used to provide an attractive document with tables and graphs supporting the text.

With the preceding in mind, there are several common types of graphs that can be used in technical documents, including (Pfaffenberger 1992):

- *Area Graph* — A line graph in which the area below the line is filled in to emphasize the change in volume from one time period to the next. The *x*-axis (categories axis) is the horizontal axis, and the *y*-axis (values axis) is the vertical axis (p. 43) (see Figure 4.L.1).

Figure 4.L.1 3D Area Chart

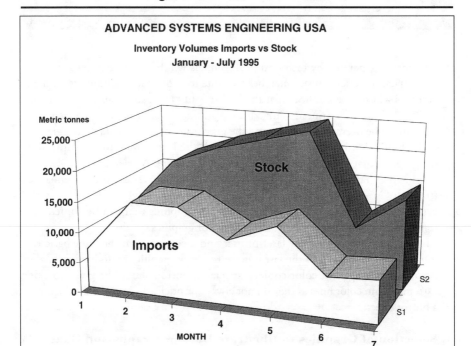

Figure 4.L.2.A Bar Chart

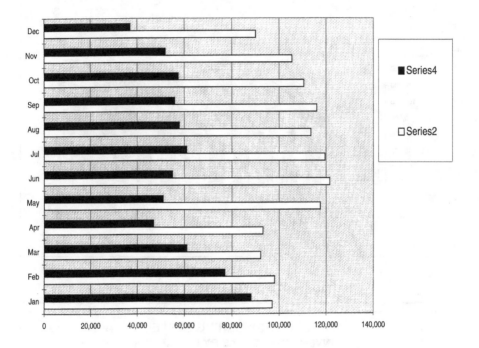

ADVANCED SYSTEMS ENGINEERING USA
DEPLETION OF INVENTORY BY TYPE BY MONTH --FISCAL YEAR 1995

- *Bar Graph* — "A graph with horizontal bars, commonly used to show the value of independent items. The *x*-axis (categories axis) is the vertical axis, and the *y*-axis (value axis) is the horizontal axis.

  Properly, the term *bar graph* is used only for graphs with horizontal bars. If the bars are vertical, the graph is a column graph . . . In practice, however, bar graph is used for both. In professional presentations graphics, bar graphs are used to display the values of discrete items (such as apples, oranges, grapefruit, and papaya), and column graphs are used to show the changes in one or more items over time. (e.g., apples versus oranges in January, February, March, etc.) (p. 66).

  Figure 45.L.2.A illustrates a standard bar chart or graph, Figure 4.L.2.B illustrates a column bar chart.

- *Doughnut Chart* — A chart which is "similar to the pie. The main

Figure 4.L.2.B 3D Cloumn Bar Chart

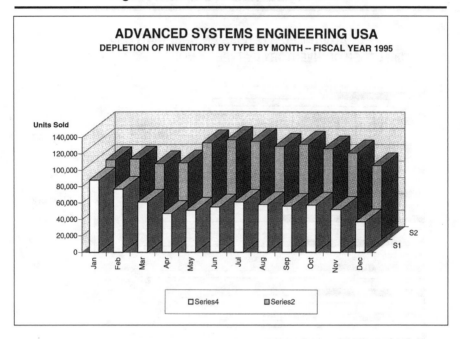

Figure 4.L.3 Doughnut Chart

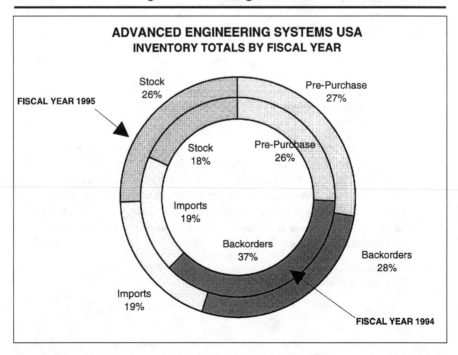

Figure 4.L.4 3D Line Chart

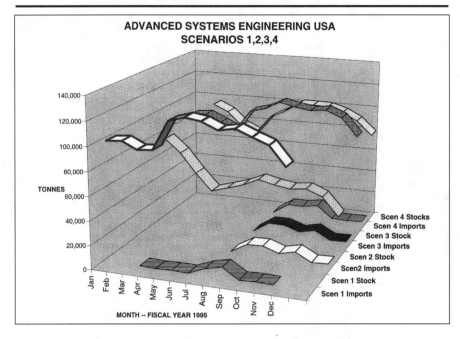

Figure 4.L.5 3D Pie Chart

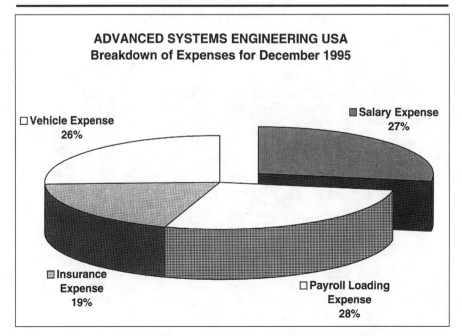

Figure 4.L.6 Radar Chart

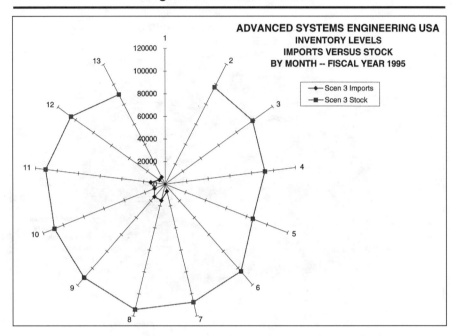

Figure 4.L.7 XY Scatter Chart

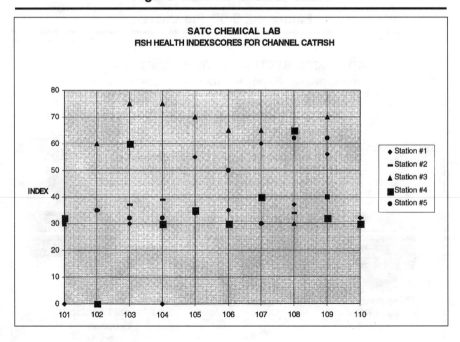

difference, other than the 'doughnut hole,' is that it can show more than one data series, unlike the pie chart. The doughnut chart is widely used in the Far East" (Microsoft 1993–94, 292) (see Figure 4.L.3).

- *Line Chart* — A graph that uses lines to show the variations of data over time or to show the relationship between two numeric variables. In general, the x-axis (categories axis) is aligned horizontally, and the y-axis (values axis) is aligned vertically. A line graph, however, may have two y-axes" (Pfaffenberger 1992, 351) (see Figure 4.L.4).

- *Pie Chart* — "A graph that displays a data series as a circle to emphasize the relative contribution of each data item to the whole.
Each slice of the pie appears in a distinctive pattern, which can produce Moiré distortions if you juxtapose too many patterns. Some programs can produce paired pie graphs that display two data series. For presentations, exploding a slice from the whole pie is a useful technique to add emphasis. (Pfaffenberger (1992), 463-464) (see Figure 4.L.5).

- *Radar Chart* — A chart that "shows changes or frequencies of data series relative to a center point to one another. Each category has its own value axis radiating from the center point. Lines connect all the data markers in the same series. The radar chart is widely used in the Far East" (Microsoft 1993-1994, 293) (see Figure 4.L.6).

- *XY (Scatter) Chart* — A chart that "shows the relationship or degree of relationship between the numeric values in several chart data series, or plots two groups of numbers as one series of xy coordinates. The xy (scatter) chart shows uneven intervals, or clusters, of data. It is commonly used for scientific data" (Microsoft 1993-1994, 293) (see Figure 4.L.7).

As with other graphics, some planning and test printing will be required for all graphs and charts. Specific attention must be given to the use of gray scales or shading in the bars, lines, pie slices, and legends. Again, not all shades or gray scales will copy exactly as printed. If a color laser printer and color copier will not be used, then a different type of filling must be used. In the case of using fillings, it is important to note that all laser printers interpret and print such shadings differently. Thus, several test prints of the graphs and charts should be run, as well as several copies made, to ensure that you get the patterns, resolution, and look so that the graph or chart information can be analyzed by the difference between each element or data series.

During the testing phase, attention should be given to the sizes and types of fonts used for titles, labels, and legends in charts. It is important to note that what may be readable on the screen may not be readable in print. A good rule to follow is to not use any font sizes smaller than 9 points in charts. Also, if you plan to reduce the graph or chart to a smaller size in the technical document, then you may have to use larger font sizes. Again, test printing and copying is the best practice to follow. These test runs should be conducted early rather than waiting until the last minute.

Figure 4.M Sample Table With Shading

3. Materials

DESCRIPTION	QUANTITY	TOTAL ESTIMATED COST
UTP Category 5 Cable (BB stock #EYN739A-R)	2,000 FT	$270
RJ-45 Plugs for solid wire (BB stock #FM110)	65	$65
3 Com Ethernet Card with 10BaseT (RJ45) connector (Entre part #3C509B-TP) ($112 each)	23	$2,576
Stackable Minihubs for 10BaseT (RJ45) (BB stock #LE2801A ($199 each)	4	$796
Conner tapes for tape drive ($39 each)	6	$234
Universal Crimp Tool (BB stock #FT046A)	1	$125
TOTAL..		$4,066

5. Selection of Graphics to Illustrate Data — Tables. Tables are another data presentation format not used as often as they could be to help highlight technical information. Tables are ideal to illustrate data in two and three dimensions. Data in four dimensions also can be illustrated, provided the numbers are small. As with graphs and charts, it is important not to use fonts smaller than 9 points. A sample table is provided in Figure 4.M, which shows a good balance of both text and data. Also, Figure 4.M shows the use of lines to help separate data elements, with additional separators of shading used. The use of shading in tables is very effective, but caution should be used as some of the shadings will not reproduce very well on a standard copy machine. A trial run of prints and copies should be conducted when using shading in tables.

Figures 4.N.1 through 4.N.9 show the screens for Microsoft Word 6.0 for Windows, which has an automated table generator called *Table Wizard.* The use of *Table Wizard* cannot only automate the table-making process but, because it is intuitive in nature, it can help produce sophisticated tables that are extremely pleasing to the eye and professional looking. However, one of the hardest things to do when using *Table Wizard* is to let the computer do the work for you, which incorporates a total of six steps.

Finally, as with graphics, graphs, and charts, explanatory text should accompany all tables. Under no circumstances is it proper to place a table in a technical document without explanatory text to highlight the important information in the table. The information in the table should be supplementary to the textual information. The table title should be short and descriptive of the information illustrated. Also, it is important to use labels and headings to properly describe the data elements. Readers should not have to guess about the information provided in the tables. Finally, refer to the respective style manual to determine the placement of table numbers, notes, and source references.

6. Selection of Graphics to Illustrate Data — Figures and Other Graphics. Like charts, graphs, and tables, the use of figures and other graphics (e.g., photographs, illustrations, drawings, etc.) is used to highlight text information and provide the reader with another format to help enhance comprehension of the material. The types and number of figures used will depend on several factors, including the target audience, technical content, and type of technical document. A training manual, for example, would generally include many photographs, drawings, and illustrations to support the training material. On the other hand, a report written for a technical audience may include many tables or formulas (see next section).

Figures should be used only when necessary. Specifically, figures should be used only if they can help clarify the text or present concepts or details that would otherwise require a lengthy discussion. There are seven recommendations when using figures, as follows:

Figure 4.N.1 Microsoft Word Table Wizard — Screen 1

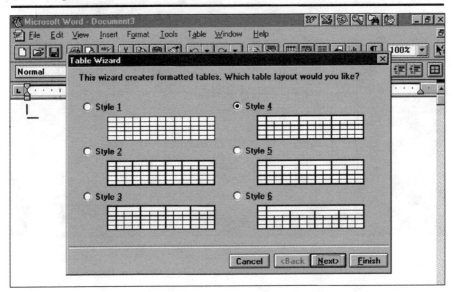

Figure 4.N.2 Microsoft Word Table Wizard — Screen 2

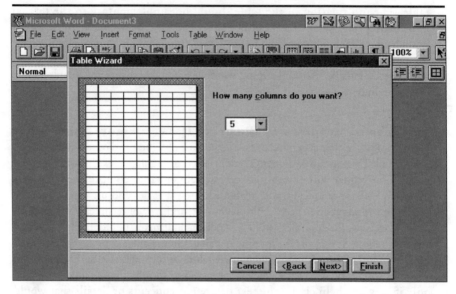

Figure 4.N.3 Microsoft Word Table Wizard — Screen 3

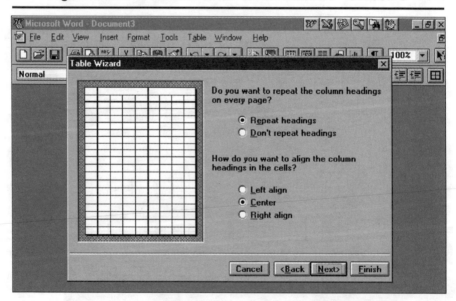

Figure 4.N.4 Microsoft Word Table Wizard — Screen 4

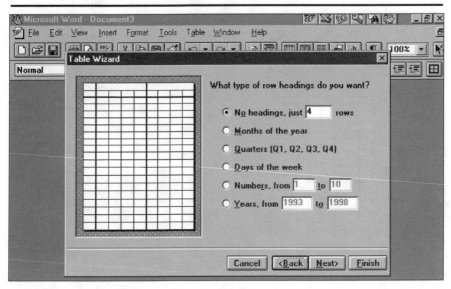

Figure 4.N.5 Microsoft Word Table Wizard — Screen 5

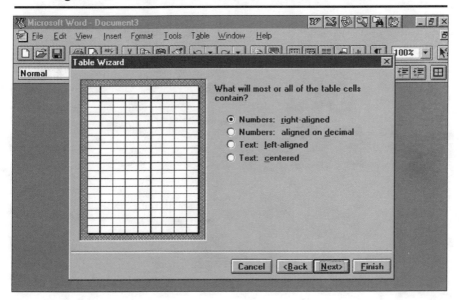

Figure 4.N.6 Microsoft Word Table Wizard — Screen 6

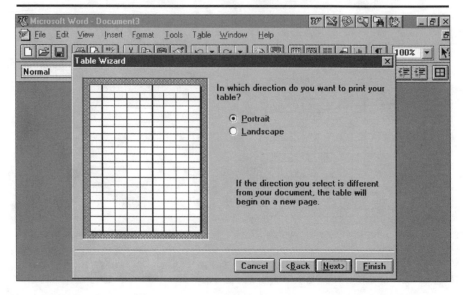

Figure 4.N.7 Microsoft Word Table Wizard — Screen 7

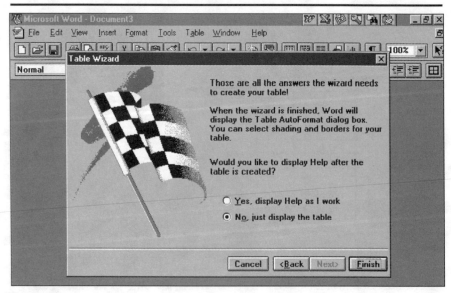

Figure 4.N.8 Microsoft Word Table Wizard — Screen 8

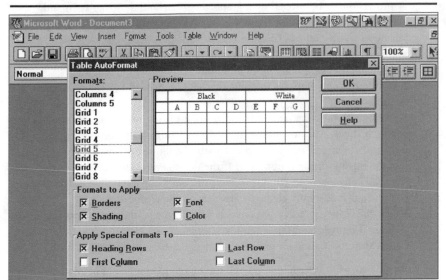

Figure 4.N.9 Microsoft Word Table Wizard — Screen 9

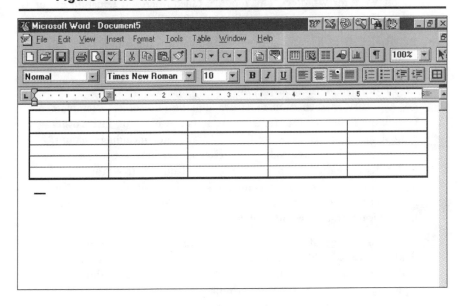

1. Use only figures that help support or clarify textual information.
2. All figures used should be generated with a computer software program or professionally developed by a graphic artist. Hand drawings (pen and pencil drawings) should be used only in emergency situations for reports, but never in procedures, training manuals, or proposals.
3. When developing figures, keep in mind what size they will be in the document (see No. 8, "Sizing of Graphics").
4. Avoid developing graphics in color that will be used in black-and-white documents.
5. If the figures must be scanned, try to use a setting of at least 600 dots per inch (dpi). Also, do not wait to the last minute to scan and place figures. Plan ahead and experiment with the figures to ensure that the highest-quality image is obtained. If the document is duplicated using a standard copy machine, conduct some print and copy tests with all figures to ensure that all figure elements are clear in the reproductions.
6. When possible, use heavy lines and boldface type so these elements will be clear if figure reduction is required.
7. Follow the style manual for placement, titling, labeling, and source citations.

7. **Selection of Graphics to Illustrate Data — Formulas and Mathematical Computations.** Thanks to the arrival of high-end word-processing programs, creation of sophisticated formulas and computations is as easy as typing a few numbers. An example of such software elements is included with Microsoft Word for Windows. This module, called Microsoft Equation Editor, is best described in the user's manual (Microsoft 1991, 2):

> Equation Editor has an intuitive, graphical interface and powerful automatic formatting that you can tailor to meet your needs.
>
> *Easy-to-use Templates and Symbols*
> For each basic mathematical construct, Equation Editor has a template containing symbols and various empty slots in which you insert other symbols and type text. There are about 120 templates, including fractions, radicals, sums, integrals, products, matrixes, and various types of brackets and braces. Related templates are grouped on palettes. You create equations simply by choosing templates and filling in their slots. You can inset templates in the plots of other templates to build complex hierarchical formulas in a natural way.
>
> Equation Editor also has palettes for inserting more than 150 mathematical symbols, several of which are unique to Equation Editor and not available in standard symbol fonts. To insert a symbol in an equation, you simply choose it from a palette — you don't need to remember any font or key combination.

Figure 4.0.1 Skewed Graphic — Example 1

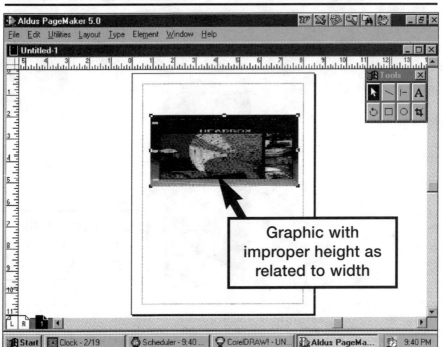

Graphic with improper height as related to width

Intelligent formatting

Equation Editor automatically handles the sizing, spacing, and positioning of symbols based on the conventions of mathematical typesetting, although you can make manual adjustments. Radicals and parentheses automatically expand or contract to fit their contents, subscripts and superscripts are sized down, and appropriate spaces are inserted around mathematical operators and relational symbols. To change the default sizes and positions of symbols, you can use commands on the format and size menus.

Equation Editor also recognizes standard mathematical abbreviations such as lim, log, and sin and sets them in the appropriate font. So, for example, if you type h(x)+ksinx, Equation Editor italicizes h, k, and x but does not italicize sin, the parentheses, and the plus sign. It inserts thick spaces around the plus sign and thin spaces around the sin function, with this effect:

$$h(x) + k\sin\ x$$

This frees you from worrying about many of the details of equation layout, improving the consistency of your work and saving you time.

Figure 4.0.2 Skewed Graphic — Example 2

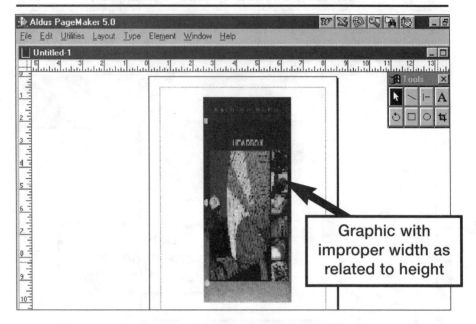

Graphic with improper width as related to height

Figure 4.0.3 Properly Proportioned Graphic

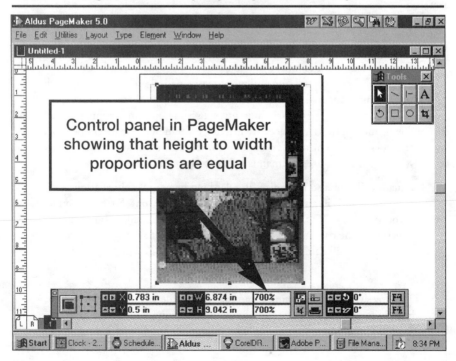

Control panel in PageMaker showing that height to width proportions are equal

Figure 4.P.1 Corel CAPTURE™ Screen

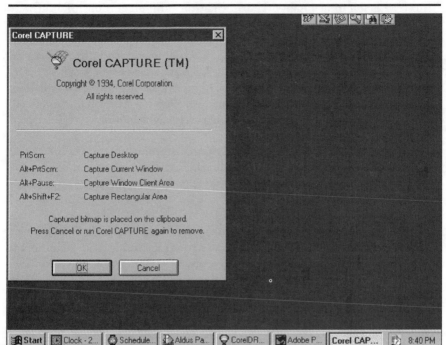

The use of programs similar to Equation Editor means that engineers, scientists, and technicians can now have full control over the development of sophisticated technical documents. With these types of programs, changes or corrections can be made with a simple double-click of the mouse. Like other technical document processes discussed within this publication, however, it is best to start experimenting with programs like Equation Editor early in the technical document development stages and not wait until the last minute.

8. **Sizing of Graphics.** Technical professionals working in a Microsoft Windows environment have the capability to size graphics placed in technical documents. The term sizing refers to shrinking the graphic so that it can be properly integrated into the document in an aesthetically pleasing manner. Sizing is usually proportional, with the length-to-width proportions exactly the same in the reduced version as in the original graphic. Because many computer users are not fully aware of the fact that in many applications programs sizing of the graphic does not always ensure that the height-to-width proportions are maintained, the sized graphics become skewed or out of proportion as reflected in Figures 4.O.1 and 4.O.2. The best advice is to read the respective users manual to determine whether the graphic can be resized while maintaining proportions.

An example of such a capability is with PageMaker, which uses a control panel to size a graph to maintain original height-to-width proportions. If a software package does not have the capability to ensure that proportions are maintained, then care should be taken so that the height-to-width proportions are maintained visually. Following such a procedure, it is good practice to print the original graphic and measure the height-to-width ratio, and then size the graphic visually and print the newly sized graphic. The newly sized graphic should then be measured to determine if the height-to-width ratios have been maintained. Adjustments can then be made to either the height or width, as illustrated in Figure 4.O.3.

Ensuring that the graphic is properly sized ascertains that it has the proper appearance and helps add to a professional look. More important, improperly sized graphics are distracting to the reader and project a sloppy and unprofessional image.

9. **Capturing Images From the Computer Screen.** Thanks to modern technologies used in many graphics programs, users can actually take a picture of the computer screen and place the picture in any Microsoft Windows application. This process is best facilitated through programs such as Corel Capture™, which now comes as a standard element with CorelDRAW™. Also, there are several shareware programs that can be downloaded from on-line services such as CompuServe that are almost as good as the CorelDRAW™ version. As shown in Figure 4.P.1, Corel Capture™ provides several options for computer-screen capturing ranging from the entire screen to the desktop portion of the screen — all

Figure 4.P.2 CorelDRAW™ Graphic Export Screen

of which are done through the selection of a keystroke or combination of keystrokes. The process of capturing an image is quite simple. First, load Corel Capture™ by simply clicking on the Corel Capture™ icon in the Corel5 Workgroup. Next, open the respective Windows-based program and file, which could be a graphics program or even a more sophisticated program such as AutoCAD. Use the keystrokes for the desired image type (full screen, desktop, etc.). Then place (i.e., paste) the image in CorelDRAW™ and save to the desired file format (e.g., EPS, PCX, etc.), along with a resolution of at least 300 dpi for clarity as shown in Figure 4.P.2. Then place the file in the desired word-processing or desktop-publishing document. The graphic can be sized or even cropped to fit the desired area. Cropping refers to cutting off from view undesired portions of the graphic.

The use of programs like Corel Capture™ is extremely useful and allows users to place almost any type of graphic from one program to another without having to convert files through bulky file-conversion routines. The process is quick and simple and can be used to place all graphics, including spreadsheets, databases, drawings, and illustrations. The only limitation is that the captured image must fit the computer screen. Images that are larger than a computer screen must be placed by more conventional processes such as scanning and importing of files.

Backing Up Computer Files During the Writing Process

It is extremely important to develop a system and set of procedures for backing up the computer files during the writing process. These precautions can protect the company against three extremely possible situations. First, the computer hard drive could fail; this type of situation happens daily to hundreds of computers and could happen to yours (this includes failure of a hard-disk drive on a network). Next is the possibility of human error, such as an employee accidentally deleting a critical project file. Finally, an individual could sabotage a project by purposely deleting all project files. This might be done by a disgruntled employee or other party who does not want to see the company secure additional business.

There are a number of good computer back-up systems that include both the tape drives and software. Once installed, backups should be conducted regularly. An example would be to conduct a total backup of the computer files weekly and a daily backup of files that have been changed, added, or modified since the last total backup. Today's back-up software programs incorporate programs that will track and conduct these critical operations automatically. The back-up tapes should be kept in a secure place such as a safe or locked file cabinet that is in a secure location. Also, it is a good idea to periodically conduct a total system backup and store the tapes off-site or at another office.

Also, when developing large and important technical documents such as proposals, training manuals, or procedures, several additional actions should be taken. First, during the early stages of writing a large technical document, it is best to do a complete backup on all project-related computer files daily. The back-up files should be kept in a secure location with limited access. Next, during the last week or stage of the writing process — a time when the writing and word-processing work is very intense — project-related files should be backed up twice a day.

Although these additional precautions seem a little extreme, they could very well protect the company against a disaster. There have been instances when a company has been in the last week, often called the home stretch of the writing process, and the computer handling most of the word-processing load has failed. Without having backed up files twice a day as recommended, a company could lose one or even several days by having to reprocess work up to the last backup. A large amount of editing has taken place during this time, based on the input of several professionals, and not all of the paper copies of editions (red lines) will be kept up with. Having a backup of the computer files twice a day means that the most that would be lost in the case of a computer failure is half a day. Going back and reconstructing editing changes and additions for a half-day is much simpler than going back several days. Ideally, the backup could be conducted during the lunch hour (unattended backup with a tape back-up system) and after work, which means that there would be no loss of writing and editing time.

Editing the
Technical Document

Although a minimum of one editor is required, having more than one competent professional read the document during the editing stages is extremely helpful. By using more than one editor, the likelihood of identifying and resolving spelling and grammatical mistakes, as well as clarifying context, is significantly enhanced. Even with the use of two editors, there is still a likelihood that some spelling and grammar mistakes will slip by unnoticed. But the incidences of such mistakes slipping by two editors are less likely.

Introduction

Editing is a formal, and essential, step in the development of all technical documents. A little extra time and effort to ensure that the editing step is properly completed will make the difference between producing a mediocre technical document and an exceptional one. Thus, it behooves every technical professional to review the two basic phases of editing technical documents. The first phase incorporates a self-editing process conducted by the author or authors. The second phase incorporates an editing process by an experienced writer who is well versed in proper grammar, spelling, and writing styles.

Essentially, there are eight reasons why technical writers must conduct the editing process properly. These reasons include the following:

1. Identify and resolve all spelling errors.
2. Identify and resolve all grammatical errors.
3. Ensure that the document is written in a clear and readable (i.e., for the target audience) fashion.
4. Ensure that the technical content is accurate.
5. Ensure that all quoted material is correct.
6. Ensure that all formulas, calculations, and other numerical data are correct.

7. Ensure that all references are correct.
8. Ensure that the document is laid out in a logical and attractive format.

The remaining portions of this chapter will focus on recommendations that should be followed in the editing process to ensure that the final version of the document is in accordance with the eight reasons for editing.

Editing by Author and Editing by Colleagues

The first logical step in the editing process is for the author or authors to edit the document. The editing process should be done in accordance with the recommendations provided in the remaining portions of this chapter. This process generally would be conducted on two or more drafts of the document before a draft is submitted to a colleague for editing. The main goal of the author's editing process is to produce a completed draft of the document which, to the best of the author's abilities, contains no known spelling mistakes (including typing errors), is grammatically correct, technically accurate, clearly written, and professionally attractive. In other words, the author's editing process should result in a high-quality draft of the document.

The next step is to have a similar process conducted by a colleague. This critical step is unfortunately skipped by many technical professionals. The more experienced writers know, however, that because they are so intimately involved with the document, they may overlook common mistakes because their subconscious knows what the document should be saying, so they do not see what it actually says and overlook simple spelling and grammatical mistakes. Equally important is the fact that because they are so intimately involved with the document and the associated project or study, they may make erroneous conclusions or recommendations based on their "desire" rather than on fact. Thus, having a colleague edit the document before printing the final version can help ensure that the document is free of errors in spelling, grammar, and even technical content.

With respect to the preceding, there are several characteristics that are desirable for a colleague who may serve as editor, including:

- The editor must be competent in spelling, grammar, and associated technical writing practices.
- The editor must be trustworthy so that the confidentiality of the technical document material is ensured.
- The editor must be reliable, that is, the editing should be completed in a timely manner or when promised.
- Although not required, it is helpful if the editor is familiar with the technical content of the document.
- The editor must have knowledge and experience in working with document layout and the elements required to develop a professional looking document.

Although a minimum of one editor is required, having more than one competent professional read the document during the editing stages is extremely helpful. By using more than one editor, the likelihood of identifying and resolving spelling and grammatical mistakes and also clarifying textual content is significantly enhanced. The incidences of errors slipping by unnoticed are less likely with two editors. Generally, such mistakes will "jump out" at the author after the document has been completed, compiled, and distributed. Unless these mistakes are critical and career threatening, it is best to simply ignore them and move on to other projects. There does come a point where the technical professional must finish the project. Sending out corrected documents due to minor spelling and grammatical mistakes after the original has already been distributed is not only unnecessary, but also draws undue attention to the respective technical professional. If the technical professional follows all of the recommendations in this book, including using one or more competent editors, chances are that the document completed, copied, and distributed will be better than those by most colleagues.

Proofreaders' Marks

Over the years a set of symbols or marks has been developed for the editing stage. The purpose of these proofreaders' marks is to provide a common set of instructions related to the editing process. Essentially, proofreaders' marks form a language no different from COBOL or FORTRAN programming languages, where symbols and combinations of symbols and text form a standardized set of instructions. From these instructions, authors, proofreaders, editors, typesetters, and word-processing personnel can easily and accurately interpret the cryptic instructions provided by key personnel working on the technical document.

The proofreaders' marks are divided into three groups: (1) general, (2) punctuation, and (3) typographical. All three of these groups, along with the respective marks and their meanings, are provided in Figure 5.A.

It is important that every technical professional learn and use proofreaders' marks because of their universal acceptance. Professionals can express suggested document changes and enhancements in a way that is simple and does not clutter up the document with a large amount of longhand writing, which can be easily misinterpreted by others. It is strongly recommended that proofreaders' marks be used by both the author and others during the proofreading and editing stages.

It is best to print the first few drafts of the document at a global space setting of two lines between text. The additional spacing will allow inclusion of proofreaders' marks and also insertion of corrections and additions to the text in a format that is easy to read at the keyboard where the changes are input. Traditionally, the proofreaders' marks and simple corrections should be made in the left or right margin of the page. Also, proofreaders should use a thin or fine-point pen with red ink. Felt-tipped pens should not be used because they easily smear. Any extended text required, such as textual corrections, should be written clearly and legibly. In some

Figure 5.A Proofreaders' Marks

OPERATIONAL SIGNS		TYPOGRAPHICAL AND PUNCTUATION SIGNS	
SIGN	MEANING	SIGN	MEANING
℈	Delete	*itl*	Set it in italic type
⌒	Close up; delete space	*rom*	Set in roman type
⌒℈	Delete and close up (use only when deleting letters *within* a word)	*bf*	Set in boldface type
stet	Let it stand	*lc*	Set in lowercase
#	Insert space	*caps*	Set in capital letters
eq #	Make space between words equal; make space between lines equal	*sc*	Set in small capitals
hr #	Insert hair space	*wf*	Wrong font; set in correct type
ls	Letterspace	X	Check type image; remove blemish
¶	Begin new paragraph	∨	Insert here or make superscript
⊐	Indent type one em from left or right	∧	Insert here *or* make subscript
⏌	Move right	⌃	Insert comma
⸢	Move left	⅋ ⅋	Insert apostrophe *or* single quotation mark
⊐⊏	Center	⅋ ⅋	Insert quotation marks
⊓	Move up	⊙	insert period
⊔	Move down	*set* ?	Insert question mark
fl	Flush Left	;/	Insert semicolon
fr	Flush right	⊙ or :/	Insert colon
=	Straighten type; align horizontally	=	Insert hyphen
‖	Align vertically	m̲	Insert em dash
tr	Transpose	n̲	Insert en dash
sp	Spell out	⦗⊦ or (/)	Insert parentheses

instances it may be appropriate to write extended corrections on a separate sheet of paper labeled "Item 1," "Item 2," and so on, and specify their appropriate location in the technical document by inserting a statement with a caret (i.e., "∧" which signifies an insertion point) labeled "Insert Item 1 here."

Proofing Galleys and Blue Lines

In the event that the document was printed at a print shop and the "proofs" are printer galleys, blue lines, or other printer proofs, the use of proofreaders' marks must be strictly followed. Also, if the proofs are blue lines or proofs made from a physical plate or negative process, then extreme caution should be taken to ensure that corrections do not result in pagination changes, which may mean additional charges. Ideally, all the changes should occur before the document is submitted to the printer or typesetter for processing, and only critical changes or typographical errors made during the typesetting process will be corrected at this stage. Also, now many typesetters prefer to receive a copy of the computer disk containing the word-processed document and will make the required conversion to their own system. Such a process will save a great deal of money, but it also makes it difficult for you to claim that all corrections submitted were typographical errors made during the typesetting process.

Spelling and Grammatical Errors

Chapter 4 contains detailed recommendations on how to ensure that all spelling and grammatical mistakes are identified and resolved. It is important to follow these recommendations while editing. It is helpful to give all editors specific written instructions to be alert to spelling and grammatical mistakes and also to alert them to the type of errors the author usually makes. Most technical professionals will have problems with certain technical writing elements. For example, some have problems with being consistent in their use of verb tense, while others may have trouble with subject-verb agreement. Sharing such weaknesses will alert editors to related problems during the editing process.

Clarifying the Text

During the editing process care should be taken to ensure that the text is written in a clear and understandable fashion. Readers should not have to guess about the exact meaning of the text. Editing for clarity means that the document is reviewed to ensure that concrete words, appropriate verbs, and only commonly accepted expressions or phrases are used. Clarity also means that the sentence structure is simple enough to get the message across — that is, no unnecessary words, as discussed in Chapter 3. For example, saying "at this point in time" instead of simply saying "now." Also, be on the lookout for the use of general terms rather than specific ones: for example, the use of the general phrase "statistical test" rather than the specific use of "conducted a regression analysis of the data." In the case of the latter, the reader does not have to guess which statistical test was used.

One proven way to ensure the text is clear is to read the document aloud. Generally, reading aloud helps identify words that add little to meaning and helps identify thoughts that should be fine-tuned to enhance or clarify meaning. Although some technical professionals may initially feel a little silly reading the document aloud, they will soon learn that the process works and helps them produce more readable documents.

Enhancing Order and Logic

One of the nice things about word processing is that nothing is in concrete. So a mistake can be easily rectified and large blocks of text can be moved from one point in a document to another. Because of the ease in moving text blocks, enhancing the order and logic of a document is much easier than in the days of the typewriter. Remember that although the order or placement of specific information seemed logical during the document development stage, it may not be as logical once the entire document is completed. Thus, it is important to consider the specific order and logic of a document during the editing process.

In some instances blocks of text may have to be moved from the beginning of a document to the end or vice versa. The main point to remember is that during the editing process it is perfectly acceptable to move large blocks of text from one point to another in the interest of clarification and logic. The only requirement is that once a move has been made, appropriate transitional sentences should be added to the material where it is placed, and any related text before and after the point where the text was eliminated should not rely on or refer to the moved text block.

Ensuring That Document Layout Is Correct and Consistent

During the technical document development stage, it is easy to lose track of the styles that were used. This is especially true when multiple style sheets were used with different level headings. As a result, chances are quite good that at some place in the document the styles used will be out of sync. Thus, it is extremely important to pay close attention to the styles used throughout the document, which includes such subtle differences as headings, subheadings, proper placement of block quotes and lists, as well as proper indentation of text. Not reviewing these elements could mean the reader will notice inconsistencies in style and may become confused about why certain styles were used in some situations and not in others. The result will be that the reader will become distracted and lose the full meaning of important information.

Additionally, extreme care should be taken to ensure that the formats used throughout the document are logical and that special styles (e.g., lists, subheadings, etc.) are used in a way that highlights important information. During the editing process, often material that should be provided in a list or under a separate heading or subheading will be identified. Although such changes may have a significant impact on the document layout (e.g., an addition of a subheading may result in having to renumber other subheadings, etc.), the changes must be made if they help highlight important information for the reader. Such changes should be made even if they require drastic changes to other portions of the document.

Reviewing Accuracy of Quoted Material

It is a good practice to go back and review quoted material, word for word, to make sure it is exactly as presented in the original document. The technical professional does not have license to change the quoted material even if there are mistakes. In the case of a mistake in the quoted material, the word *sic* should be placed in brackets [sic] immediately following the incorrect material in the quote.

Reviewing Correctness of Tables, Formulas, and Other Data

Other areas that require scrutiny during the editing process are tables, formulas, and other data (hereafter called "data"). Sometimes the data is entered in an incomplete or rough format during the document development stage with the idea that it will be double-checked or finalized during the editing stage. Unfortunately many technical professionals are rushed during the editing stage and fail to double-check critical data that support the document thesis. The result is that due to transposed numbers and other typing mistakes, the data are incorrect. In other instances the technical professional fails to double-check addition, subtraction, multiplication, division, or other operations that result in incorrect totals. In both situations the professional reputation, and perhaps the job or contract, of the technical professional will be severely compromised.

The best way to avoid such potentially career-damaging situations is to implement a two-step data review process during the editing stage. The first step is to carefully double-check every document element that contains data. During this process special care should be taken to ensure that all data are correct according to the source document or documents and that all data operations are computed and presented accurately. This means that hand calculations should be conducted for all tables presented in the document. This first review step should be accomplished before the document is fully read by the author. All corrections or changes should be noted in red ink in the data elements.

The next step is to review all textual references to the data. In some instances the data may be changed in a table due to updating, and the technical professional will forget to change the corresponding data reflected in the text. Again, as discussed here, all data presented in tables, graphs, and related figures should be discussed in the text of the document. Obviously, if the data in the text and that presented in tables, graphs, and other figures do not agree, the reader will not be impressed and will most likely question all conclusions or recommendations provided in the document.

Use of Copyrighted Material

Copyright laws have been in effect in one form or another since 1710, when "the English Parliament passed the first copyright statute, known as the *Statute of Anne*" (Samuelson 1995, 15). Since then many countries, including the United States of America, have adopted their own version of copyright laws. According to *Circular 1: Copyright Basics* published by the Copyright Office, Library of Congress (1994, 2–3), a copyright is a:

> Form of protection provided by the laws of the United States (Title 17, U.S. Code) to the authors of "original works of authorship" including literary, dramatic, musical, artistic, and certain other intellectual works. This protection is available to both published and unpublished works. Section 106 of the Copyright Act generally gives the owner of the copyright the exclusive right to do and to authorize others to do the following:
>
> - *To reproduce* the copyrighted work in copies or phonorecords;
> - To prepare *derivative works* based upon the copyrighted work;
> - *To distribute copies or phonorecords* of the copyrighted work to the public by sale or other transfer of ownership, or by rental, lease, or lending;
> - *To perform the copyrighted work publicly,* in the case of the literary, musical, dramatic, and choreographic works, pantomimes, and motion pictures and other audiovisual works . . .
>
> . . . Copyright protection exists from the time the work is created in fixed form; that is, it is an incident of the process of authorship. The copyright in the work of authorship *immediately* becomes the property of the author who created it. Only the author or those deriving their rights through the author can rightfully claim copyright.
>
> In the case of works made for hire, the employer and not the employee is preemptively considered the author. Section 101 of the copyright statute defines a "work made for hire" as:
>
> 1. a work prepared by an employee within the scope of his or her employment; or
> 2. a work specially ordered or commissioned for use as a contribution to a collective work, as a part of a motion picture or other audiovisual work, as a translation, as a supplementary work, as a compilation, as an instructional text, as a test, as answer material for a test, or as an atlas, if the parties expressly agree in a written instrument signed by them that the work shall be considered a work made for hire . . .
>
> The authors of a joint work are co-owners of the copyright in the work, unless there is an agreement to the contrary.

Copyright in each separate contribution to a periodical or other collective work is distinct from copyright in collective work as a whole and rests initially with the author of the contribution.

Finally, categories of material that generally are not covered by the Copyright Act include (Office of Copyright 1994, 3):

- Works that have *not* been fixed in a tangible form of expression. For example: choreographic works that have not been notated or recorded, or improvisational speeches or performances that have not been written or recorded.
- Titles, names, short phrases, and slogans; familiar symbols or designs; mere variations of typographic ornamentation, lettering, or coloring; mere listings of ingredients or contents.
- Ideas, procedures, methods, systems, processes, concepts, principles, discoveries, or devices, as distinguished from a description, explanation, or illustration.
- Works consisting *entirely* of information that is common property and containing no original authorship. For example, standard calendars, height and weight charts, tape measures and rulers, and lists or tables taken from public documents or other common sources.

Contrary to popular belief, a document does not have to be published or filed with the Copyright Office of the Library of Congress in order for it to be considered "copyrighted material." Under the provisions of the Copyright Act, a document is automatically copyrighted when it is created when "it is fixed in a copy or phonorecord for the first time" (Office of Copyright 1994, 3). For purposes of the Copyright Act, "Copies are material objects from which a work can be read or visually perceived either directly or with the aid of a machine or device, such as books, manuscripts, sheet music, film, videotape, or microfilm"(office of Copyright, 1994). Of course, registering a document with the Copyright Office is strongly recommended as such proof can be extremely persuasive in any legal actions you may undertake in the event that another individual duplicates or distributes unauthorized copies of your document.

The preceding excerpts of the Copyright Act have several important implications for technical professionals involved in developing technical documents. First, the law specifically protects the copyright of materials created by others, which means the technical professional must be fully aware of the permissible uses of copyrighted materials referenced, cited, or otherwise adopted in technical documents. Second, unless otherwise stipulated in writing before the development of a technical document, the technical professional's employer will hold the copyright on all materials developed by the professional as a part of his or her job responsibilities.

When using copyrighted material, the technical professional must be cautious and follow the provisions of the Copyright Act, which may mean obtaining permission to use the material in the document and give the original author or authors appropriate credit. Generally, such a request can be facilitated through an application for permission to use copyrighted material as shown in Figure 5.B. The application should include such critical information as the applicant's name and address, specification of the excerpt or material that will be used, title of the document that will include the copyrighted material, a specification that the applicant will agree to give the original author appropriate credit, and a place for both the applicant and original author to sign for certification. For published documents, the application can be sent to the publishers; for private works, the application can be sent directly to the original author.

Additional information on copyrights regulations can be obtained from:

> Copyright Office
> LM-401
> Library of Congress, Madison Building
> 101 Independence Avenue
> Washington, DC 20559-6000
> (202) 707-3000 (TTY: 707-6737)

The best recommendation is that permission to use copyrighted material be obtained for works that will be or have the potential to be distributed to more than a few employees within the company. This recommendation prevails even though there is a widely used provision of the Copyright Act called the "Fair Use Doctrine." Basically, this doctrine was developed for situations involving "scholarship," such as at educational and research institutions. The premise was that most of these institutions are nonprofit organizations and thus have a limited need for use of the copyrighted material, which is a small amount of the original material duplicated specifically for classroom use. Of course factors such as the amount of original material copied, the number of copies, and the frequency of use are considered in the Fair Use Doctrine.

The Fair Use Doctrine applies only if the use is extremely limited and will not be harmful to the rights of the copyright holder. The phrase "will not be harmful to the rights of the copyright holder" generally means that the copyright holder will not lose a substantial amount of money due to fewer sales.

Unfortunately, beyond the scholarship situations, the applicability of the Fair Use Doctrine to the private for-profit sector is doubtful. Thus, it is best to seek permission for the use of copyrighted material in technical documents — especially for technical documents that will be distributed to more than a few employees. In either situation, proper credit must always be given for works or parts of works created by others.

Figure 5.B Sample Form for Obtaining Permission to Use Previously Published Materials

345 East 47th Street, M/S 6B, New York, NY 10017 • Telephone (212) 705 7075 • Fax (212) 705 7841

Request for Permission to Reprint Previously Published Materials

Date: _____

To *(material copyright holder)*: _____

From *(author requesting permission)*: _____

    Address: _____ Phone: _____

    City/State/Zip: _____ Fax: _____

I hereby request permission to reprint the following material from your publication:

    Title of periodical or book in which material appears: _____

    Author(s)/Editor(s) of periodical or book in which material appears *(if any)*: _____

    Title of article or chapter in which material appears: _____

    Author(s) (of article/chapter or book): _____

    Date of Publication: _____ Vol. and No. (if applicable): _____

    Type of material to be reused (circle one): Figure   Table   Illustration   Text   Other _____

    FIGURE (number and page:): _____ ILLUSTRATION (number and page): _____

    TABLE (number and page): _____ OTHER (identification and page): _____

    TEXT (beginning and ending phrases and page(s)): _____

This material is to appear as originally published (or with changes/deletions as noted on the reverse side of this letter or in attachment) in a work currently being prepared for publication by ASME Press. This permission is for world rights for all editions, and for any advertising, promotional, or publicity matter or purpose (including dust jacket or cover) related to the publication and marketing by ASME of the following publication:

    Title of ASME Press book: _____

    Author/Editor of ASME Press book: _____

    Title of article or chapter of ASME Press Book in which material will appear: _____

    Author of article or chapter in which material will appear: _____

    Estimated publication date: _____

Approval of Request by Copyright Holder

As the copyright holder of this material, this request for permission to use material is hereby approved, provided the following credit line is used: _____

Signature of Copyright Holder: _____ Date: _____

Copyright Holder: Please return this form to the author requesting reprint permission at his/her address listed above.

ELECTRONIC DOCUMENTS: FACSIMILES (FAXES) AND ELECTRONIC MAIL (E-MAIL)

"E-mail is the foundation of electronic communication. It breaks down the barriers of time and place, and has particular benefits for employees who are frequently away from the office — on the road, working at home or in satellite facilities. In addition, E-mail brings a business closer to what Alvin Toffler refers to as "adhocrocy." In this model, the typical corporate information hierarchy is broken down, and communications flow freely from senior management to factory floor workers, and everywhere in between" (Rudnick et al. 1995, 11).

M. Rudnick, S. Wiener, and J. Kaplowitz
Employee Communications and Technology:
The_revolution.begins@now

Introduction to Faxes

Ten years ago a publication such as this would not even have a footnote reference to fax technologies for use in business and industry. The fax explosion of the late 1980s has surely changed the manner in which business is conducted. As reflected in Figure 6.A, it is estimated that currently there are more than 10 million fax machines being used in business and industry, plus another 6 million fax cards installed in personal computers. In total, there are well over 16 million machines that have the capability of receiving and sending faxes over standard telephone lines. Also, as reflected in Figure 6.A, the number of fax units (fax machines and fax cards) will grow to more than 46 million units by the year 2000 (Vanston et al. 1991, xv).

It is almost impossible to visualize how business could be conducted in today's "instant" world without faxes. And, as illustrated in Figure 6.A, the usage of fax units in business and industry, as well as a virtual explosion in the use of home fax units, will continue to increase and become an even more critical element in the lives of employees and private citizens. Given the rapid expansion of fax unit use, one would think that the technology was relatively new.

Figure 6.A Fax Machines and Fax Cards in Use

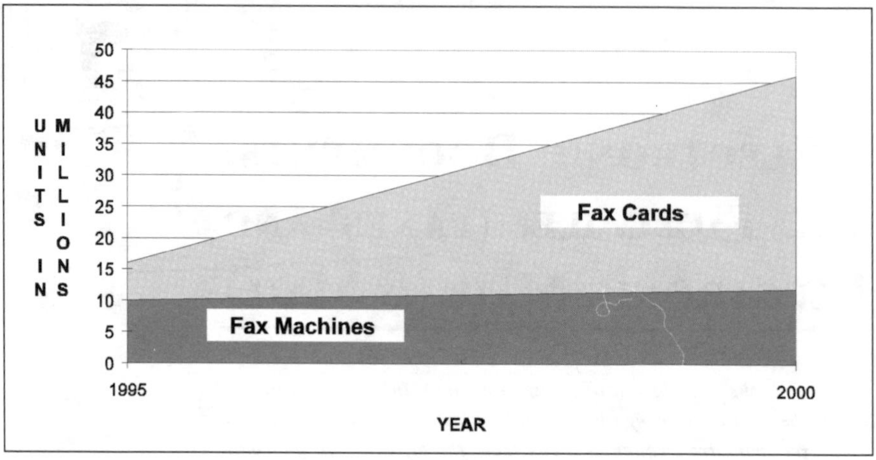

Actually, fax technology dates back to 1843 when the first patent was awarded to a British clockmaker named Alexander Bain (Coopersmith 1993, 46).[11] The Bain technology incorporated raised metal blocks and a stylist that was not only crude but extremely expensive to create. Like all technology, it went through several improvements and refinements from 1865 to the 1940s, when war divided the world. Due to World War II and the urgent need to transmit maps, weather charts, enemy troop movements, and battle plans, both the German and Allied forces used fax technology, which included expending research and development efforts to improve the quality of fax technology. By the late 1960s use of fax technology expanded to a larger commercial basis, with approximately 25,000 units in use by 1970. Then fax technology was seen as a strong potential market for business and industry, with several manufacturers beginning to design, produce, and sell fax machines. The real growth of the fax market was in the 1980s when less offensive "smelling" paper was used and microtechnology helped to place the purchase of the fax machines at a price comparable to a professional-level calculator. It soon become a permanent fixture in every office and plant.

Even with rapidly changing technology, fax machine units will continue to be an integral part of how business is conducted at work and in the home. In fact, current estimates show that the business and industry fax unit market is almost saturated, which is precisely why the fax industry is looking to the home fax market as its next frontier. According to Technology Futures, which has conducted several critical telecommunications need and impact studies on telecommunications, there will be almost 96 million home fax units in use by the year 2010 (Vanston et al. 1991, xv). The reasons for the broad acceptance and popularity of fax technology are many and significant. A sampling of these reasons includes:

---

[11]Portions reprinted with permission from *IEEE Spectrum,* Volume 30, Issue 2, p. 46. Copyright ©1993 by IEEE.

- Cost of fax units is extremely reasonable; a good fax machine sells for as little as $300 and quality, high-speed fax cards for computers sell for around $150.
- Fax technology is becoming an alternative to regular mail and allows the instantaneous transmission of important business documents. Many news stories show that fax technology is now in direct competition with the U.S. Postal Service. This competition is the result of perceptions of poor service by the Postal Service and the fact that when you compare long-distance telephone rates and first-class postage rates, it is now cheaper to send faxes than mail at the first-class postage rate.
- A printed document is more accurate than the spoken word over the telephone and is less susceptible to misinterpretation. As result, the transmission of a document provides a written record (documentation) of project activities and agreements.
- Faxes can be transmitted 24 hours a day to any location in the world where there is a telephone line, cellular phone network, or organization that has access to a satellite telephone system. Time zones and datelines are no longer a barrier to conducting business.
- More courts are now accepting faxes as legal documentation, although proof of fax transmission may still be a weak link. Proof that a fax has been received is easy to obtain from a printout of the fax machine or fax card's internal log, but the log printout cannot prove that the fax has actually been received or given to the intended recipient. There is no doubt that fax technology manufacturers, lawyers, and court systems will eventually devise a certification process or technology to provide legal documentation that the intended recipient has received the fax.
- Faxing allows engineers, scientists, and other technicians to receive and review engineering documents and drawings. A timely review may be critical because of equipment failure or other emergencies and probable causes can be discussed over the telephone. Without the immediate transmission of these engineering documents and drawings, an engineer would have to be dispatched to the site or a long delay would be incurred due to overnight mail service. Such delays could cost millions of dollars in downtime or even lives in emergency situations.

Using Fax Technology

Most of today's fax units are Group 3 and "conform to the standards for Group 3 devices, as set in 1981 by Study Group XIV of the International Telegraph and Telephone Consultative Committee (CCITT)" (Vanston et al. 1991, 11). The specifications of the Group 3 fax unit are as follows:

Transmit the digital data at 9600 bits per second (b/s), with each bit representing a 0 or 1. Normal resolution for Group 3 fax is about 200 x 100 dots per inch (dpi). This means a scan across the page is made 100 times for every inch of travel down the page. On each scan across the page the image is sampled (stored as a 0 if white or 1 if black) 200 times for every inch of travel across the page. Each square inch of page requires about 200 x 100 = 20,000 bits. Thus, an 8-1/2 - x 11 - inch page requires about 8.5 x 11 x 20,000 = 1,870,000 bits. At 9600 b/s, a page would take 1,870,000/9600 = 195 seconds to transmit. (Vanston et al. 1991, 12)

For most business documents, Group 3 fax units will be sufficient. Resolution and resulting readability problems could arise during the transmission of drawings and photographs.

Group 4 fax units are currently under development. The Group 4 fax technology will take full advantage of the newer telecommunications networks such as Integrated Services Digital Network (ISDN), which can transmit "data at 56 to 128 kbps (kilobits per second) — that is two to four times the speed of v.34, the fastest modem standard" (Brisbin 1995, 105). Through ISDN-type networks, fax transmission can take full advantage of transmission rates of 56 to 128 kbps, which will allow transmission of higher-resolution images (400 x 400 dpi). Prior to the use of ISDN networks, high-resolution transmissions were not practical due to lengthy transmission time required over regular telephone lines where speeds are limited. The additional transmission time would tie up telephone lines for unreasonable amounts of time and cost a fortune in long-distance rates. Current estimates show that initially the Group 4 fax machines will cost $10,000 per unit but will, as all new technology does, come down in cost as more manufacturers enter the market and production costs due to economies of scale and competition force the prices to a universally acceptable cost.

Given the projected high cost and requirements for ISDN or telecommunications networks, most organizations will continue to use the Group 3 fax technology for the next few years. But given the low, and universally accepted, cost of Group 3 technology, there is no reason why systems below Group 3, which are slower and have lower resolutions, should even be used. It is highly recommended that organizations purchase a Group 3 unit. Almost all Group 3 units provide for a high or fine, and even superfine, transmission setting. Of course the higher-resolution settings will take longer to transmit than the standard resolution settings. If pictures, graphs, tables, or drawings are being transmitted, the superfine or highest-resolution setting should be used.

No matter which type of fax unit you use, you should use a cover sheet. Given the high volume of fax traffic on today's telecommunications networks, which has caused an explosion in new telephone installations and helped "deplete the pool of available local telephone numbers, hastening the split of area codes" (Vanston et al. 1991, 1) a fax cover sheet is intended as a privacy measure for all fax users. The average fax machine in today's workplace serves approximately 12 employees (Vanston et al. 1991, 13). As a result many faxes are received by a given unit, and the cover sheet helps ensure that the fax is given to the right person. Whatever the design or layout of a fax cover sheet, the following critical elements should be included:

- *Name, Company, Address, and Telephone Number of the Fax Recipient:* The name and address of the recipient is important to ensure that other employees using the recipient's machine know where to forward the fax. If the fax is accidentally sent to the wrong fax number (which happens often), individuals at the receiving end will know who to contact to let the receiver know that a fax was inadvertently sent to the wrong number. Most of the fax units used today have speed dials with preprogrammed telephone numbers of clients, customers, and others who are sent faxes regularly. The only drawback to using speed dials is that it is extremely easy to push the wrong speed dial button and have the fax sent to the wrong party. Although the cover sheet will not necessarily prevent the wrong recipient from reading the fax, it will give the individual enough information to telephone the intended recipient.

- *Total Number of Fax Pages Transmitted:* Sometimes the fax machine will receive the entire fax and store the information in a buffer while printing. Also, if it is out of paper, some fax machines will store the information in the buffer until additional paper is loaded. Finally, most of the software used with fax boards allows the computer user to store all faxes for later printing or print as the fax is received. Most fax units will indicate that all pages were successfully transmitted. This is no guarantee that the recipient's fax machine can print all pages. Power failures, surges, and other unforeseen problems can occur just as the last page is printing, and the information in the buffer will likely be lost. Specifying the number of pages transmitted on the cover sheet will alert the recipient to check that all pages have been received.

- *Name, Address, and Telephone Number of Sender:* A recipient may be working on a project and receiving a number of faxes on the subject matter. Providing this information on the cover page will be helpful as the recipient organizes subject information. Also, most recipients should keep the cover sheet with the fax as the sender's information, including any special notations, will be helpful as the recipient reviews the project research and information sources.

A little-known fact is that federal laws mandate that the name and telephone number of the fax sender be included with every fax transmission. Specifically, 47 CFR § 68.318(c)(3) states that:

It shall be unlawful for any person within the United States to use a computer or other electronic device to send any message via a telephone facsimile machine unless such message clearly contains, on the margin at the top or bottom of each transmitted page or on the first page of the transmission, the date and time it is sent and an identification of the business, other entity, or individual sending the message and the telephone number of the sending machine or of such business, other entity, or individual. (Fed. Reg. 1992, 48336)

Although most fax units in use today do print the sender organization, telephone number, date, and time of transmission, there is no guarantee that the person who set up the fax unit bothered to program it with this information. To ensure that the required information is reflected on the fax, it is a good practice to simply use a cover sheet.

- *Date and Time of Fax Transmission:* The legal rationale for including this information on the fax transmission has already been reviewed. Also important, however, is the rationale that this information helps the recipient determine the specific order in which documents were received. Such a determination can be very critical when different versions of a document are being transmitted after changes or editing.
- *Disclaimer:* About the only thing a disclaimer statement will do is keep honest people honest. Including a disclaimer statement will "remind" individuals that a fax addressed to others is not for public reading, duplication, or distribution. The disclaimer statement should be at the bottom of the fax cover sheet. A recommended disclaimer statement is as follows:

> The information contained in this facsimile message is intended only for the individual or entity to whom it is addressed and may contain information that is legally privileged, confidential and exempt from disclosure under applicable law. If the reader of this message is not the intended recipient, you are hereby notified that any dissemination, distribution, or copying of this communication is strictly prohibited. If you have received this communications in error, please notify us immediately by telephone and return the original message at the above address via the U.S. Postal Service. Thank you.

The following actions can help automate and enhance the fax transmission process to ensure that such documents are properly transmitted, to project a professional image of the sender, and to discourage unprofessional curiosity.:

- Design and development of a quality fax cover sheet that includes the company logo and information already discussed. Additional helpful information can include an e-mail address as well as normal working hours. The final version of the fax cover sheet should be of high resolution and high quality. Preprinted fax cover sheets can be used so that only the addressee information needs to be written in, along with the number of pages (including the cover sheet) being transmitted. Print a large quantity of the cover sheets on a quality laser printer. Do not keep making copies of copies of the cover sheets; they look unprofessional and sometimes cannot even be read.

A Few Words on Fax Etiquette

Miss Manners® on Fax Etiquette

By Judith Martin

DEAR MISS MANNERS — Exactly how public is a fax message that is sent to my office and addressed to me?

Is it like a postcard (which my mother and the mail carrier brazenly and openly read), or is it like a business letter which only I or my secretary may open?

A colleague approached me waving my fax — concerning a very sensitive company matter — reading it aloud, unapologetically, and pontificating about what my response ought to be. She has since demanded to know the outcome of this matter, which has no bearing on her job whatsoever.

What to do? Move my desk right beside the fax machine?

GENTLE READER — Postal-card rules indeed apply to the fax, but Miss Manners notices that you have the misfortune to be surrounded by people who do not practice postal-card etiquette. The rule is that no one must read an open communication, but that no one must assume that everyone else has not read it.

Got that?

In other words, no strictly private message should be sent unsealed. And yet it is a transgression of etiquette for anyone to acknowledge having read anyone else's mail, no matter how openly it was sent. So your colleague was polite in delivering the fax, rude in reading it aloud, and unspeakable in inquiring about it. Your response should have been "I believe that was intended for me." As to your letter carrier, Miss Manners cannot imagine why he hasn't died of boredom long ago from reading the inane messages people send from their vacations.

© 1993 United Feature Syndicate, Inc.

- Use of more recent fax software packages with computer fax boards that have the capability of overlaying a custom-made fax cover sheet in the cover sheet sent with the fax. Such customization allows the user to include recipient name, address, company, fax telephone number, telephone number, as well as date and time of transmission. Such templates also allow inclusion of the sender's name, address, company, fax and telephone numbers, company logo, and any other information. One such software package is Smartcom Fax produced by Hayes in Atlanta, GA. Although development of the template may turn out to be quite a chore, the result is that each fax sent using the fax board includes a professional-looking cover sheet that is completely automated. All the user has to do is select the name or names from the software phone book of the recipient to send a fax with all of the legally required information. A copy of the fax cover sheet designed with the Smartcom Fax software is included in the samples at the end of this chapter. A sample cover sheet developed for a fax software program is provided in Figure 6.B

- Checking of the fax unit's transmission parameters to make sure that the quality of the fax is at least on fine rather than standard. In the case of transmitting photos, graphs, charts, or other graphics, the setting should be set to superfine or the highest resolution.

- Checking of the fax unit's transmission parameters to ensure that a log of the transmission is either stored in the unit's memory or a copy is printed after the transmission. Although such documentation is not proof that the fax was actually received by the intended recipient, it will show that you attempted to send the fax. Such proof is enough to document "good faith" effort.

- Checking of the fax unit's transmission parameters to ensure that the fax is stored in memory when recording paper runs out. This is extremely important as most fax units are not cared for or maintained as they should

Figure 6.B Sample Fax Cover Sheet From TechWrite, Inc.

From (334) 575-4760   to 13138865679   at 03/03/96 02:25p   Pg 001/004

FAX TRANSMITTAL FORM

TechWrite, Inc.
320 Perry Avenue
Monroeville, AL 36460

Telephone #:   (334) 575-7396
Fax #:   (334) 575-4760

Internet E-mail
Address: 72430.60@COMPUSERVE.COM

TO: Fred Belonge

COMPANY: TCA Engineering

DATE: 03/03/96   TIME: 02:25p

TELEPHONE #:   1-313-884-9645

FROM: John H. White, Ed.D.

NUMBER OF PAGES TRANSMITTED: _____ 4 _____

If you do not receive all the pages transmitted, PLEASE CALL (334) 575-7396 AS SOON AS POSSIBLE.

NOTE OR MESSAGE:

Fred:

Attached are the requested equipment specifications. Call me
if you have any questions or require additional information.

John H. White

be. Usually the fax unit is in a central location and in continual use. Employees send and receive faxes as needed but pay little attention to see whether the unit has paper. If the memory capability is not activated and the unit runs out of paper, only the fax portion that was printed on available paper will be printed — the rest of the fax transmission will be lost. On the other hand, if the memory is activated, then the remaining portion of the fax will be stored in the memory and an alarm will sound or light up informing users that the unit requires paper and that there is still information stored in the memory. When paper is added, the unit will print the remaining portion of the fax.

Finally, before sending any fax, several concerns should be addressed to ascertain that the fax (cover sheet and attached document) is transmitted in a way that ensures the rights and privacy of all concerned as well as protects the proprietary rights of your company, clients, or customers. Also, consideration should be given to the quality of the fax and your professional image. These concerns are as follows:

- Fax transmissions may not be the correct forum for distributing several types of documents. First, consideration must be given to security and protecting the rights of others. Faxes on sensitive personnel matters, should only be transmitted to a fax machine that has limited access. Information such as employee evaluations, employment status, termination notices, and reprimands should not be sent by a fax. These are issues that must be addressed face-to-face.

- Fax transmissions such as medical records should be sent only to a secure fax machine with limited access.

- Documents such as bid quotes and proposals that you are working on for your company should not be sent by fax. Because of low security and central locations of most fax units, anyone walking by the unit could see the information or even take the fax. If, because of time constraints, bid quotes and proposals must be faxed, send the document to a fax unit that has limited use. Also, in highly competitive industries dealing with contracts in the millions of dollars, encrypted fax units (in pairs) should be purchased for each of your company offices to ensure that only the designated fax unit can receive and access the fax transmission. Costs of these encrypted-type fax units are very reasonable; they may pay for themselves with only one fax transmission. Not only do such units ensure that individuals tapping telephone lines will not have assess to the fax information, but it also means that faxes accidentally sent to the wrong telephone number will not be received.

- Always use a fax cover sheet. Do not use the little fax notes that paste on the first page of the documents. These look unprofessional and may cover up information on the document. Also, because they are so small and the resolution of the fax is limited, often you can't even read the names of the sender or recipient. If you use these little notes, your fax may never reach the addressee.

- Keep in mind that in addition to your name and company name being printed on the fax cover sheet, the name of your company or branch location may be programmed in the fax unit. As a result, your company's name will be printed on the top margin of every page faxed. There will be those few instances where you send a fax to someone with a properly completed fax cover page and ask the recipient not to tell anyone where the fax came from. With your company's name or branch location printed on the top margin of every page, it will not be difficult to determine where the fax originated even if the document is circulated without the cover page.

- Although fax technology has many wonderful uses and provides many benefits to society, it does have its limitations. Remember that the quality of the fax document received will depend on the quality of the fax transmitted. Faxes that contain many colors, especially blues, will not transmit clearly using the standard Group 3 fax units. If in doubt try sending the fax to yourself through another fax machine near your work area. If you only have one fax machine, try making a copy of the document using a setting a little on the light side. The copy will illustrate whether all colors are being picked up.

- Font size used in the document can also be problematic. Actually, documents with most font sizes smaller than 10 points per inch will not be fully legible on the transmitted document. There are two simple ways to resolve this problem. If it is a word-processed document, simply raise the point size to 12 or 14 points per inch, print a new copy, and then fax. Make a copy of the document on a quality copy machine that has an enlargement capability. The enlarged copy can then be faxed with all of the document fully legible. Also, use the enlarger on a copy machine when small charts, maps, diagrams, or other graphic-intensive documents are to be faxed.

- All information on the fax cover sheet, including notes or comments, should be typed or word-processed whenever possible. Not only will the fax look more professional but, more important, typing will ensure that all the information is legible. This point is extremely important if your handwriting is hard to read on regular documents, let alone a fax where the resolution is lower than on the original document faxed.

- One effective way to keep snooping eyes off faxed documents lying around in another office is to use a fax board with a personal computer. Most fax board software has the option to print faxes as received or later. Unlike the memory of the fax machine, the fax board receives the fax and makes a hard copy in a computer file format (using a TIFF graphic file). Once the file is stored on the hard disk, it may be retrieved and printed on a laser printer at will. The file will still be there even if the computer is shut off after the entire fax has been received. Besides having the ability to print the fax when needed, the TIFF file can be imported into word-processing and desktop publishing documents.

Introduction to E-Mail

Although fax technology has enhanced methods of quick document transmission, e-mail is revolutionizing the way business is conducted. During the past few years, the transition to these sophisticated, internationally linked systems has moved at such a lightning pace and, according to a study conducted by Cognitive Communications and the Document Company Xerox, e-mail "is the technological application most widely used for employee communications. Nearly nine of 10 respondents (87%) said their company uses e-mail for message transmission between individuals" (Rudnick, et al. 1995, 11). Equally important, the study found that "two-thirds of the respondents said that e-mail is used to disseminate publications."

Given the current entrenchment and growth of e-mail, it is safe to say that this new form of workplace communications will be around for many years. Unfortunately, like other new forms of technology in the workplace, very few employees have the experience or training to effectively use this new medium that has become the foundation of corporate communication. This lack of experience and training can cost a company thousands or even millions of dollars due to miscommunication, breach of confidentiality of proprietary information, and even liability due to an employee's misuse of a company's e-mail system.

Given the potential liability associated with lack of experience or training in the use of e-mail systems, this chapter focuses on four elements to help the technical professional fully understand the strengths and weaknesses of this communications technology as well as effectively use these systems as a conduit for the transmission of technical documents. These elements include:

1. Issues dealing with confidentiality of e-mail messages and documents
2. Etiquette of e-mail
3. Writing in e-mail style
4. Mechanics of e-mail and developing large documents for e-mail transmissions

The remaining portions of this chapter deal with these four e-mail elements. Also, the last section includes several e-mail samples and formats from actual e-mail documents and messages.

Mechanics of E-Mail

Ideally there are three configurations of e-mail systems: (1) internal corporate e-mail systems, (2) external e-mail service systems (MCI Mail, Sprint Mail, AT&T Easylink, etc.), and (3) internal corporate e-mail systems that are also linked with a gateway to an external e-mail service system, which can include the internet.

The first type of e-mail system, internal corporate, is the easiest to use because all address conventions are usually the same and there is a system administrator who can be easily contacted for help. Also, internal e-mail messages are usually delivered instantaneously.

The second type of e-mail system is a little more difficult to understand and use. These systems use a modem, communications server, mail server, or other form of gateway to link with the main system hub so the message can be processed and routed through the Internet. Naturally, the time delay between transmission and receipt of the e-mail message through the external system will be a little longer.

The last type of system is a combination of internal corporate e-mail systems that are linked with external e-mail service systems. Because these systems are bulky and difficult to use, many companies are eliminating these systems for new ones that use the Internet for both external and internal messages. An example of such a system will have a multitasking mail server that uses a program called SL Mail for the server and mail client, such as EUDORA Pro, for the workstation interface. With such a configuration, the company staff would have Internet e-mail addresses using, as a minimum, the Post Office Protocol 3 (POP3) standard. Also, the same addressing convention would be used to send messages to other Internet users. Such a configuartion is cost-effective and easy to use. The only downside of such configurations is that they are not as secure as the internal system. But if security for internal e-mail messages is critical, then encryption software can be purchased at a modest cost.

Whatever the actual type of e-mail systems used, all three have several of the following common elements:

- Provision to specify who the e-mail should be addressed to
- Provision to specify who sent the e-mail
- Provision for an e-mail subject
- Ability to send copies of the e-mail to users besides the addressee
- More than enough space for the e-mail message
- Ability to select a "reply" option for received e-mail
- Ability to forward a copy of a received e-mail message to another user
- Ability to attach a computer file to the e-mail

With internal e-mail systems, almost all forms of computer files can be transmitted through e-mail. On external systems or when transmitting through an external system, there will be some limitations on the type of files that can be attached. Some systems, for example, allow for the transmission of almost any type of file from a graphics file to an electronic spreadsheet. On the other hand, some systems allow only for the transmission of text or ASCII files. Extreme caution should be taken when attaching files to e-mail messages, especially if you are trying to send the file through an external system. In such cases there are several steps that should be taken:

1. Check with the addressee to see what types of files their system will accept as attachments.
2. Have the addressee ask the system administrator about the maximum file size that can be received.
3. Check with your system administrator to make sure that you can transmit your files through the associated gateway and if there are any file size limitations.
4. Make a small test file in the same program in which the intended file was developed. The test file should then be attached to an e-mail message

asking the addressee to open the file and make sure that it was transmitted without any damage to the file.

5. Do not send attachment files that involve extremely confidential or competitive information — there is always a chance that the file will have problems getting through the entire gateway and will end up at the terminal of the e-mail system administrator for return. If the message gets damaged, the administrator may even open the file to determine where to return the file.

Most systems also include a provision that e-mail will be returned in the event that it cannot be delivered either because of an incorrect address or a system failure. With an external system, which includes linking with external systems, it may take an hour or two before the e-mail is returned.

Issues Dealing With Confidentiality of E-Mail Messages and Files

One of the first things that new users of e-mail must comprehend is that of all current communication systems used by professionals, e-mail is an extremely insecure system. Computer network system administrators, for example, generally have unlimited access to each e-mail account and its contents. Additionally, depending on the size and type of organization, supervisors may also have full and unlimited access to the e-mail accounts of employees under their charge.

Because of the relative newness (compared to the telephone) of e-mail, relevant laws have not been fully tested in the courts. The result is that many "experts" claim one or two things in terms of e-mail user rights. First, some of the more liberal experts claim that employers should not have free and unlimited access to an employee's e-mail — arguing that to permit full and unlimited access violates an employee's constitutional rights to freedom of speech. Second, some of the more conservative experts, who are generally on the side of business and industry, claim that access to e-mail is just a new type of employer right that falls under the umbrella of intellectual property in the workplace. They claim because the e-mail was developed on company time and equipment and transmitted or received on company equipment, it is property of the company which means the supervisor has full and unlimited rights to access an employee's e-mail account. This includes access to the employee's personal e-mail.

During the past few years there have been several landmark court cases dealing with the privacy of employees and the rights of employers to read their employees' e-mail. One case involved Nissan and Rhonda Hall and Bonita Bourke, who were hired to establish and manage the company's e-mail system. Once they had the system up and running, a "female supervisor heard that some of their e-mail was getting pretty steamy and began monitoring the messages. She soon discovered that the two had some disparaging things to say about her, and the women were threatened with dismissal.

When Hall and Bourke filed a grievance complaining that their privacy had been violated, they were fired"[12] (Elmer-Dewitt 1993, 46). Nissan lawyers successfully argued that since the e-mail messages were produced on company computers, the supervisor was within her right to read any and all messages on the system.

The Nissan case and similar others show that the courts have been leaning toward employer rights over employee rights in regard to e-mail transmissions. The main reason for this is the newness of e-mail in the workplace and the lack of laws that itemize specific employee rights of e-mail privacy. In fact, the only federal legislation that even deals with electronic communication is the *Electronic Communications Privacy Act (ECPA) of 1986*. "Under the *ECPA*, it is unlawful for a person or entity furnishing electronic communication service to the public to knowingly disclose to a third party the contents of a communication that is in electronic storage, carried, or maintained by the service" (Shieh and Ballard 1994, 60). The operative provision in the *ECPA*, however, is "furnishing electronic communication service to the public." There is little doubt that this legislation was designed to cover telephone and message firms that provide service to the public and with a legal implication, which appears to be supported by the courts, that the law was never intended to protect the privacy of employees in the workplace. In an effort to provide some form of electronic privacy to workers, Representative Pat Wilson introduced the *Privacy for Consumers and Workers Act (PCWA) of 1993* (H.R. 1900). The *PCWA*, however, was never acted upon and died a quiet death.

Given the lack of clear legislation explicitly protecting employees' privacy in the workplace regarding access to their e-mail and the employers' strong record of winning such cases, it is safe to say that at least for the next few years, employees should be aware that their e-mail may be legally read by their supervisors. The safest course of action is to be cautious about what you say in your e-mail and keep such messages on a professional level. Personal messages or messages that libel someone have no place in the work environment. Employers can help the situation by developing clear corporate policies dealing with access to e-mail messages. If supervisors are going to read employees' e-mail, then the employees should be told when and under what circumstances their e-mail may be read by a supervisor or other employee. Informing the employees that their e-mail may be read is expected. Equally important, if employees know that their e-mail may be read, they will most likely limit their correspondence to a professional nature, which can save all parties anger, embarrassment, and legal costs.

E-Mail Etiquette

Newcomers to e-mail systems are always puzzled about the fact that these electronic messages are written in short and sometimes cryptic formats. Although they can usually read and understand these messages, they do not understand why they are

---

[12] Copyright © 1993 Time Inc. Reprinted by permission.

not written in a traditional letter or memorandum format. The best way to fully understand why e-mail does not follow traditional conventions is to take a quick look at the history of e-mail. When Internet was in its formative years and known as ARPAnet, electronic messages were sent between mainframe computers that were large and kept in deep-freeze temperature rooms with platform floors for the many bulky cables linking the system's work stations and printers. These systems were owned or leased by universities, research centers, or military facilities that had limited computing resources. Also, these systems were often run by serious computer users (the politically correct name for computer "nerds") who spent most of their waking hours writing and debugging programs. Through the ARPAnet linkages, system users could send electronic messages as well as "talk" with users logged on the net from other mainframe systems. Talk is a two-way, interactive communication format that allows one user to type real-time messages to a selected user or users on the system and active linked computers. Common practice in these talk sessions was to send a short message and get an equally quick response because the system talk routines would actually interrupt the users from their computing activity. Only a short line or two were sent quickly so all parties could get back to work. Also, on these older systems, talk sessions would drain valuable computer-processing resources. Along with the talk sessions, users would send quick, short e-mail messages to one another. The e-mail messages were kept short for two reasons. First, composing long-winded messages kept these serious users from their programming, debugging, and processing responsibilities. Second, due to limited user directory storage space, messages were kept short so as not to clutter or use up the space.

The next link in the evolution of e-mail came when mass production of mainframe systems increased, thanks to the microprocessor, which significantly lowered the capital outlay for the mainframe systems. The result was that more companies were able to purchase their own mainframe systems. Mass production and purchase of personal computers began in the early 1980s. With these large purchases the computer technology started to expand from the serious computer user who needed to know complex computer programming routines to the everyday user in the office or plant who, because of the newer user-friendly programs, could operate a computer without having to be a programmer.

Simple-to-use e-mail systems soon appeared on the market and offered an alternative way for employees to communicate with one another. Many of these new users, however, were trained on these systems by the original serious users who had their own way of communicating via e-mail, and picked up the "jargon." Today, this jargon is being used in business e-mail as well as on the Internet — the last frontier for the serious computer user that is now being overrun by the average employee.

Although the e-mail jargon has an interesting history and can be fun to use, the time has come to focus on developing e-mail documents and reports that can be read and understood by all readers. It is extremely important to remember that today's e-mail is an integral part of commerce. Also, e-mail is now used and read by employees at every level and position of a company or organization, and the recipient of your message could very well be a fellow employee, the chief executive officer of your company, or even a customer. In all three situations you want to make certain that you use clear and concise language that offers little or no opportunity for miscommunications that could cost you your job and cost your company large sums of money or a large contract.

With the preceding in mind, there are several rules that should be followed when using an e-mail system to help make your communication clear and represent you as a professional. These rules are as follows:

- Since most e-mail systems allow one line for a subject on the message, be concise and as descriptive as space will allow to get the attention of the addressee. Like your mail box at home, today's e-mail boxes receive high volumes of both important and junk mail. With the glut of junk e-mail, several computer software vendors have developed sophisticated e-mail packages to prioritize your e-mail. These filtering systems are best described by Adam Strack with U.S. Robotics, who had a big problem with e-mail overload. Strack states:

> The situation improved considerably in early 1993, when we got hold of cc:Mail for Windows version 2.0 . . . The new version includes a Rules feature — a rare but brilliant idea that, fortunately, more and more e-mail vendors are building into their systems.
>
> In a nutshell, Rules lets me program the mail system to perform a task if certain events occur, such as if I receive a message from my boss. I simply make a selection from the rule menu, which has a list of 17 common rules, such as "Archive messages older than February 4th" and "Notify me if I receive a message from the boss." Now when a message arrives, the system checks the sender name and subject heading. If it's my boss sending me mail, my PC sounds a bell.
>
> I've also programmed Rules to let me know if I get a note from any of my other supervisors or from a customer via Internet. In addition to the options that come standard with the program, we've built a few extras to meet our specific needs. For example, to keep private messages confidential, we've created a rule to file any correspondence marked private in the manager's mailbox, rather than forwarding it to someone else (such as an administrative assistant).
>
> The Rules-based system also enables me to filter noncritical messages. For example, I know that almost all the messages I receive from the human resources department are about a policy or benefit change, so I created a rule to file those messages into a folder named HR. (Fryer 1994, 43)

- Putting a clear subject heading on your e-mail will help increase the likelihood that your message will not get lost in junk mail or, worse yet, get put into the virtual garbage can before it is read.
- Use full and proper names for the "To" and "From" lines. Do not use

nicknames or just first names. Using the full names makes it clear from the start who is sending the message.

- Some e-mail systems can request a return receipt to verify if the e-mail was delivered to the addressee. In cases where documentation or having a strong paper trail is important, requesting the return receipt would be a good idea.

- Do not use computer words or abbreviations — use real-world words and terms. Examples of computer abbreviations are "BTW" which stands for "by the-way" and "a.k.a.," which stands for "also known as." These abbreviations may be acceptable for the hobbyist who is "net surfing" but not for the business world. Also, these words are acceptable when you are in a "talk" or "chat" mode on the Internet but have no place in business and industry communications. Using such abbreviations greatly increases the chances that the real meaning will be misunderstood.

- Think twice before transmitting an e-mail message that is extremely confidential or could help support a case against you for libel. Traditionally, e-mail messages are not secure; there are individuals such as network system administrators and staff who have complete access to your e-mail files. Thus, extremely confidential or potentially damaging personnel information is best transmitted through other channels, such as overnight delivery couriers or a face-to-face meeting.

- Only send e-mail that is important and meaningful. Cute little cryptic messages (e.g., "great idea, boss," "OK," or "right on," etc.) are a waste of time and if you, or the addressee, are using commercial e-mail systems, these messages are actually wasting money.

- Be aware of where you are sending your e-mail — especially if you are sending a message to an employee of a federal agency. There are several lawsuits working their way through the federal court system promoted by special-interest groups that are trying to make the federal government treat e-mail like all government documents. The legal basis for these suits is strong because the *Federal Records Act* requires that U.S. agencies keep copies of all important business correspondence" (Feds 1994, 255). If these groups are successful, then all e-mail sent to any federal agency will become a "public record," which means the general public must be given access if requested.

- Use initial capital letters and lowercase letters as you would in a letter or memorandum. Avoid using all capital letters; it is not only hard to read, but to serious computer users use of all capital letters signifies anger, otherwise known as "flaming."

- Be professional — do not use computer "smileys" — a computer-generated figure that represents different facial expressions: for example, a happy smiley would look something like this ":-)" and was generated by using the colon key, dash key, and close parentheses key. Using a smiley may be "cute" or acceptable in some settings, but has no place in professional messages.

The 5th Wave   By Rich Tennant

"QUICK KIDS! YOUR MOTHER'S FLAMING SOMEONE ON THE INTERNET!"

©The 5th Wave by Rich Tennant, Rockport MA. E-mail: the5wave@tiac.net

- Be careful of the tone of your message — like all writing, the tone must help convey your meaning.
- Try to avoid using underline and boldface fonts, as many older e-mail systems will not display these styles. Also, for the same reason, use simple fonts.
- Remember your intended audience and follow all of the traditional target-audience checks discussed in Chapter 2. There are a few publications in the market that encourage users not to follow the rules — following all the rules takes too much time. True, following the traditional grammar rules does take more time, but your reputation and job are at stake. Not using the rules will lead to sloppiness, which means that your message may well be misunderstood. Keep in mind that e-mail has not, as of this date, replaced the U.S. Postal Service. Most readers are tuned to reading grammatically correct correspondence, and they will notice glaring grammatical errors that are common when e-mail

users take shortcuts. What some e-mail users view as acceptable in bending the rules of grammar, the average professional may very well misinterpret as a sign that the sender has never mastered the fundamentals of writing. This is not the image that you want to give your boss, customer, or potential customer.

- Keep your sentences and paragraphs short.
- There will be times when all you have to do is send a one- or two-sentence e-mail message in response to a message or simply to ask a quick question. These types of messages are acceptable since you still follow all of the spelling and grammatical rules.
- Common courtesy dictates that you answer all business-related e-mail. Some e-mail systems allow for a "reply" response, which should be used when possible. Using the reply option maintains what is called the "thread" of the message and will tell the addressee that your message is in reply to a message that you received. The reply can even be a short "thank you," "thanks for the information," or "I will give the idea some thought and get back with you." The main point is to acknowledge receipt of the e-mail message. Sending an acknowledgment receipt or message is becoming an industry controversy as e-mail system traffic has increased on both the Internet and corporate systems. The traffic increase is due to more and more companies providing wider systems access to employees and a significant increase in junk e-mail. In companies with limited computing power, which could be the result of antiquated systems still being used or an unanticipated increase of system use, this increase has caused systems to "crash" or shut down. As a result, system administrators at these underpowered sites are demanding that only "critical" messages be sent to the outside world via e-mail. Hopefully, such limitations will only affect underpowered sites and will be lifted once new and more powerful systems have been purchased and installed. Thus, the best advice to follow is to check with your e-mail administrator and make certain there are no limitations to sending e-mail. If there are no limitations, then follow the practice of sending an acknowledgment message to the the sender.
- If you require an answer or response from the addressee, specify the deadline date in the text of your message.
- Do not send any e-mail message that is offensive in nature to an individual or group of individuals.
- Be patient, there are still many e-mail users who write their messages in computer language. If you do not understand a message, simply send a return message asking for a clarification.
- Do not send e-mail in anger — the reduced communications cycle of e-mail can get both parties angry much quicker because the "cool down" period associated with the postal service or even interoffice mail systems is significantly reduced.

Developing Large Documents for E-Mail Transmissions

There is a significant difference in sending a one- or two-line e-mail message and a larger message that may be two or more pages. The biggest difference is that you must be aware of the layout of your message and understand that what you see on your computer screen is not necessarily what your addressee will get. Because of this, you must do a little planning before you even begin to draft your document. As you plan your e-mail document, there are five basic practices that should be followed:

1. Be aware of the margins and spacing of your e-mail system and the system being used by most of your recipients. If, for example, you have large margins extending to around 85 characters per line, your message may be received in a wraparound fashion that will make it hard to read. One way to resolve this problem is to set your default margin in your e-mail system between 60 and 79 characters per line. Also, send an e-mail message to one of your colleagues using another type of e-mail system and have the colleague return a printed version of the message via fax. This copy can then be compared to your original version to see if there were any formatting problems, and adjustments can then be made accordingly so that your e-mail messages are properly formatted when sent to other e-mail systems.

2. Be aware of the number of lines that will be transmitted per e-mail page. A good rule to follow is that you can get approximately 35 lines of message to each e-mail page. Because every system differs on how it will format a received message, use paragraph or section headings sparingly.

3. Because of pagination problems, avoid using phrases referring to below or above. Pagination on the receiving end of your message may be different from that of your system and computer. Pagination changes coupled with location references may be confusing to the reader.

4. Ideally, if you have an e-mail message over two pages, it may be safer to send a short message showing the purpose of the correspondence and attach a file with the longer report or document. Most of today's high-end word processors provide for the conversion of a document to an ASCII format. The bulk of your message will be intact; the only real loss is that you will not get special character formatting (e.g., boldface type, underlining, etc.). If, however, your system and the recipient's system allow for the transfer of binary files in the original word-processing format, then most of your special formatting will be preserved (see Figure 6.C). Using an attached file also provides a little extra privacy for someone accidentally reading an e-mail message as the reader must go through several steps to open and read an attached document.

5. If you are sending a large e-mail document, it is a good idea to specify at the end of the message or file "– End of message please send return message to sender specifying that entire message has been received –." Requesting such a reply may entail extra trouble for the recipient but will help ensure that

Figure 6.C Sample E-Mail Program With
File Attachment Capability

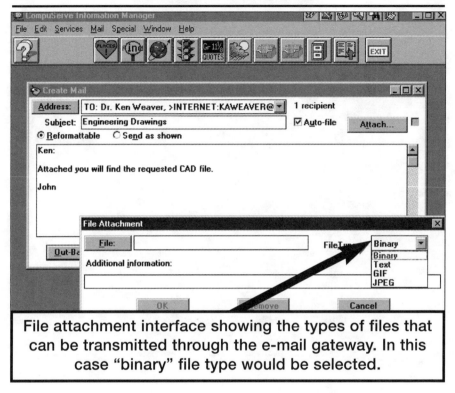

File attachment interface showing the types of files that can be transmitted through the e-mail gateway. In this case "binary" file type would be selected.

the complete message was received. If, after a reasonable period, you do not receive verification, then send another short message asking if it was received.

Taking these few extra steps will ensure that your document is received and readable and will show the recipients that you are a professional and have gone the "extra mile" for his or her benefit.

LETTERS AND
MEMORANDUMS

For memos of transmittal, commitment, confirmation, and other short topics, the body of the memo should simply include an introductory sentence, topic sentence or sentences, and closing sentences. For more complicated topics or situations where a longer memo is required, then extreme care must be taken regarding what the memo should contain. Mastering Memos, *a video from the editors of* Communications Briefings, *states that "most memos should never be written," (*Mastering Memos *1992). The point is that most of the information in memos in the workplace is meaningless or could have been more efficiently communicated verbally. Because of the "worthless memo syndrome," the technical professional must first make certain that the memo is needed to convey important and meaningful information before beginning a draft. Once it has been determined that the memo is the best format, then the next logical step is to plan your memo in terms of who, what, where and why . . .*

Introduction to Letters

Today, as in years past, the letter is still a very important document that the technical professional will be required to write. And because of its important role and its similarity to business documents, there are literally thousands of books that deal with how to write effective business letters. Unlike other forms of technical writing, most technical professionals have been exposed to formal letter writing in English courses. And with a few minor differences, letter writing for business situations is not very different from letter writing for technical situations.

Given the abundance of letter-writing publications and courses, this chapter focuses primarily on those crucial technical writing elements that are unique to the letter that the technical professional will be required to write. As a result, this chapter has been written based on the assumption that the reader will have been exposed to the fundamentals of business letter writing.

Types of Technical Letters

Although there are many types of letters, the technical professional will generally work with five types. These technical letters include:

1. *Letter of Transmittal* — One of the most common types of letters that the technical professional will use. This letter explains the purpose of an attached document and serves as documentation that the item was sent to the recipient. Good practice dictates that a letter of transmittal will be used in all instances when a document or report is being sent to another individual outside the company. The letter of transmittal should be short and to the point. Where a documented paper trail is required, the letter of transmittal should include "chain of custody" sign-off at the bottom of the letter documenting receipt. In such situations, two copies of the letter of transmittal would be sent, with one copy returned to the sender after the recipient has signed the letter of transmittal.

2. *Letter of Instruction* — This is used when transferring information or knowledge to others. The letter of instruction should be clear and concise and in accordance with all other technical writing practices. The letter of instruction should include the following:

   a. An introductory paragraph to give the purpose and background of the instructions.
   b. Bulleted or numbered lists in logical and sequential steps.
   c. Any precautions or safety measures that should be followed.

3. *Letter of Follow-up* — A follow-up letter is used after a face-to-face meeting, telephone conversation, or for confirmation or clarification of receipt of a letter or document. The follow-up letter can serve as documentation or simply as a courtesy. In most cases a follow-up letter is short and to the point. Ideally, there will be four parts to a follow-up letter:

   a. *Linkage Paragraph* — Linking the purpose of the letter with a specification meeting, telephone conversation, or receipt of an item(s) or information. Common practice would be to specify the date of the contact or receipt.
   b. *Verification* — Specifying the agreement or understanding of the information or matter.
   c. *Closure* — Ending the correspondence in a professional manner and putting the ball back in the recipient's court for further action. It is extremely important that as far as the sender is concerned, the sender has done all that is required (professionally, morally, and legally) for the matter at hand.

4. *Letter of Inquiry* — A letter of inquiry should be simple and contain only critical information. Unless you are writing a letter of inquiry on a product

that a particular company manufactures or sells, which would be handled by a marketing or sales representative, a letter of inquiry is not given a high priority in any organization. Thus, spending hours and hours in writing a letter of inquiry will be a waste of precious time. Following are three basic rules for writing letters of inquiry:

a. Give as much detail about your inquiry as needed. Provide only relevant and important information.

b. If you need the information immediately, politely indicate that you are in need of the information as soon as conveniently possible.

c. If you are addressing the letter to a specific person, you may want to state that you will call in a few days to see if there is any other information that is required. This approach is helpful when you need the information immediately or if you know that the respective party has a reputation as a procrastinator. If you say you will call in a few days, however, make sure that you call as promised. The recipient may need additional information and may simply decide to wait for your call.

5. *Letter of Application* — Recent studies have shown that the average worker will change careers as many as four times during his or her life. Such changes are prompted by drastic changes in the economy as well as the fact that relationships with employers have "deteriorated from 'til-death-do-us-part' to 'what-have-you-done-for-me-lately'"[13] (Connelly 1995, 145). Beyond the obvious fact that this statistic indictates that today's technical professional must look out for economic security (i.e., put money away for self-funded retirement), it also means that sending letters of application for a new position will be much more common than in the past 20 or more years. Pressure is added because the competition for those desired jobs (e.g., good salary and benefits, a company that is stable, a company that is progressive, etc.) is very high. Thus, getting the "attention" of those executives in the position of screening applications is critical. As previously discussed, packaging, or how the letter looks, is as important as the information provided within. Again, if you cannot get the attention of the reader, you will never make it to the interview stage. Given this dictum, the basic rules should be adhered to when writing a letter of application:

a. The letter should be professional and attractive and must be printed using a laser printer. Do not mail a letter of application using a dot-matrix or ink-jet printer.

b. Do not be afraid to brag about your qualifications.

c. If you know a *"reputable"* contact in the company — mention the name in the letter. Regardless of what people say, the "good ole boy (or gal) network" is alive and well. Do not be afraid to use those

---

[13] Connelly, J. Fortune, © 1996 Time Inc. All rights reserved.

contacts. Most professionals actually obtained their jobs through initial personal contacts (e.g., friend, professors, colleagues, and family, etc.). All this means is that the contacts will help get your letter read at least. Common curtesy dictates, however, that you have the permission of the contact *before* the letter is mailed.

d. Be sure to state what you have to offer the prospective company.

e. Unless you are applying for a high-level executive position, limit the letter to one page. Remember, the purpose of your letter of application, and the accompanying resume, is to get an interview.

Finally, there are three key parts to a letter of application:

1. *Opening Paragraph* — A short paragraph that serves as an introduction and a link with the position that you are applying for. Also, names of any company contacts should be specified in the opening paragraph.

2. *Marketing Your Skills and Experience* — The next two, but no more than three, paragraphs should provide highlights of your qualifications, including experience and professional preparation that are directly related to the position that you are applying for. Any special experience or achievements should also be specified.

3. *Open-door Closure* — A short closure paragraph should be included at the end of the letter. The closing paragraph should also specify that you welcome the opportunity for an interview. More aggressive job hunters even add a statement that they (the applicant) will call in the next week or two to talk more about their qualifications for the job. Although this last approach is risky, it has been effective in situations where the screening staff is overworked or has been screening hundreds of applications. The "follow-up" telephone call can leave a positive impression if the respective individual can use the telephone in an effective manner (i.e., persuade the other party that an interview would be worthwhile).

Technical Letter Format

Most companies have already established a letter format that must be followed for all official business-related correspondence. The format includes detailed specifications on using corporate letterhead and exact measurement guidelines on the placement of all letter elements. A graphical example of such guidelines is illustrated in Figure 7.A, which has been provided courtesy of Vanity Fair Mills. Like most organizations, Vanity Fair Mills specifies the use of a "full block" letter format. Traditionally, the full block letter format has all of the letter elements in a flush-left alignment. The full block format is simple to set up on any computer, is functional, and when combined with the correct font size and style, it presents a professional image as

Figure 7.A Standard Letter Format Template From Vanity Fair Mills

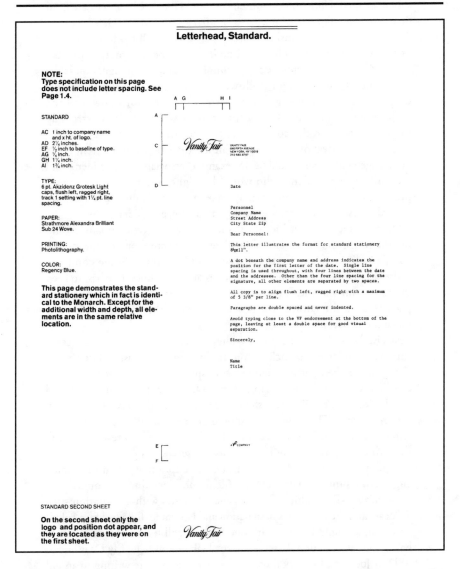

reflected in the letter in Figure 7.A.

The full block letter has a total of 10 elements for letters used with corporate letterhead. These elements include:

1. Date Line — The date should be specified in the full format such as "June 1, 1995." Some companies and organizations require the specification of the date in the military or international format, such as "1 June 1995."

Unless your company requires the international format, the standard format should be used. Under no circumstances should the date be abbreviated. An example of an abbreviated format is "03/01/95" or "MAR 03, 95."

2. *Inside Address* — The inside address specifies the full name and address of the letter recipient. The first line includes the designation of Ms., Mr., Dr., or other accepted designation. Following the designation is the full name of the recipient, followed by a comma and the respective corporate title. Do not guess the title; if you do not know, call the addressee's office and ask a receptionist or secretary. It is extremely important to have the correct title for executives of a corporation as there is a great deal of sensitivity regarding titles between different levels of management. There could be, for example, associate vice presidents and only one vice president, who would be higher than an associate vice president. A vice president may take offense if one of his or her subordinates receives letters with the title of "vice president."

The next line is reserved for the name of the respective company. Again, use the proper name; do not abbreviate or use acronyms. Following the company line are the first and second address lines. Some company addresses may be exceptionally long and require two lines. It is best to keep the address on one line. If two are required, make the line break at a logical place so as not to confuse the postal service or other couriers that must make the delivery.

Finally, the last line of the "inside address" is for the city, state, and zip code. International locations such as country, province, or postal code should also be placed here for international correspondence. Also, use the full zip code if it is available. The full zip code includes the traditional five digits and an additional four digits. This should help speed the delivery of your letter.

3. *Subject Line* — The subject line should be used for all types of technical letters. The subject line helps ensure that the recipient can readily identify the purpose of the correspondence. Also, it serves as a good reference for your filing of the letter. Along with a text description of the subject, some engineering professionals include a file number or project account.

4. *Salutation* — The salutation line generally starts with a salutation such as "dear" and the recipient's designation and last name, followed by a colon. An example would be Dear Dr. Jones:. Ideally, all technical letters should have a formal salutation. Reserve informal salutations (e.g., My Dear Friend or Dear Joe, etc.) for only those instances where you are writing to an old friend and the letter will not be distributed to other professionals.

During the past few years there has been a "trend" that when the letter is signed, the writer puts a line through the formal salutation and writes the recipient's first name in blue ink. Such practices are not professional, give the appearance of a gimmick, and are distracting to most readers. In other words, avoid trends when it comes to writing in a technical setting. This is especially true when we consider that most engineers and engineering-related corporations tend to be a little more conservative than the professionals over on Madison Avenue.

5. *Body Text* — The body text of your letter should follow all other technical writing conventions covered in other chapters that deal with writing style. The text should be single-spaced. Keep paragraphs short and use bullets or numbers to highlight important information or information that is best presented in a broken format rather than in long sentences and paragraphs. Also, remember to watch the tone of your letter.

6. *Closure* — After the text of the letter comes the closure, which is usually a formal closing such as "Sincerely." Again, reserve more informal closings (e.g., "Best wishes," "Yours truly," etc.) for old friends and when the letter will not be distributed to other professionals.

7. *Signature Block and Title* — The signature block title is reserved for the actual signature of the writer, with the full name and title on the next two lines. Be sure to leave at least three lines between the "close" and the "full name" so that there is enough room for the signature. Also, if the letter is going to be mass-produced and you wish to still add a personal touch, you can have your signature scanned and converted to a "TIFF" or "PCX" file, which can then be placed as a graphic in your letter. The signature can be placed and, if required, sized, to look as though you have signed every letter. The authenticity can be enhanced if you have access to a color laser printer so that your signature is printed in blue and the rest of the letter in black.

8. *Writer's and Typist's Initials* — Because many professionals word process their own letters, the requirement for writer's and typist's initials is not as strong as it once was. Ideally, the use of the initials is functional when the letters are being typed in secretarial pools or where one secretary may work for several technical professionals. It is, however, acceptable if the initials are left off; they do not serve any important or critical function. The important element is who signed the letter, not who was responsible for the word processing.

9. *Enclosures or Attachments* — The enclosure or attachment line is reserved for a specification of any enclosures or attachments that will accompany the letter. As with the memo, enclosures are usually for books or reports that are too large to attach to the letter or would stand on their own merits and not be used with the letter. In turn, attachments are reserved for smaller documents that not only accompany the letter, but are generally used with the letter. The best practice is to individually list the enclosures or attachments as illustrated in Figure 7.B. It will help the reader and provide you with an accurate record of what was sent. Again, such documentation may prove very helpful at a later date, such as in a legal proceeding.

10. *Computer File and Directory Name* — The listing of the computer file and directory name will be helpful when editing or retrieving the letter at a later date. You will have a letter that you may want to use again and

Figure 7.B Sample Letter With Recommended Elements

TechWrite, Inc.

May 27, 1996

Mr. Terry Sparco
President
Dynamic Test Labs, Inc.
3456 Industrial Park
Birmingham, AL 34678

Dear Mr. Sparco:

As requested, I have reviewed your telecommunications equipment needs for linking your branch offices with 64 Kbps leased lines. Based on this review, I believe that the Cisco 2525 Router with Token Ring LAN interface, 56Kbps/64Kbps DSU/CSU will meet your needs. Based on the estimated cost of $3,700 per unit, you will need a total of six units with a total cost of $22,200 not including installation and annual maintenance. For your convenience, I have enclosed a brochure from Cisco with the specifications of the 2525 Routers.

I will call you next week to arrange our next project meeting. Meanwhile, please do not hesitate to contact me if I can be of further service.

Sincerely,

Dan Wolf
Director, Technical Operations

JHW

Enclosure: Brochure on the Cisco 2525 Router

d:\winword\DTL\Tsparco1.doc

TechWrite, Inc.
320 Perry Avenue
Monroeville, AL 36460
205/575-7396

send to another employee, customer, or client. By specifying the file name and directory location, you can locate the file in a matter of minutes so that all you have to do is change the date, inside address, and salutation. This practice will save a great deal of word-processing time that can be used for more important tasks. The computer file and directory names should be placed two lines below the enclosure or attachment line. If attachments or enclosures are not used, then the computer and directory file names should be placed two lines below the writer's initials. An example of a computer file and directory names specification would be: c:\winword\docmnt\asme\dwizda.doc. In this example, the file name is "dwizda.doc" and is located in the subdirectories of "asme" and "docmnt" in the directory of "winword" on the "c" drive. Thus, any employee who has a copy of the original letter can locate the desired computer file in a matter of minutes instead of going through numbers of directories or even hundreds of floppy disks.

General Rules on Letter Writing

Basically, letter writing is a simple and straightforward process if several common-sense rules are followed:

1. The letter should be simple and to the point. A dissertation is not required and not expected when it comes to letters.
2. Sentences and paragraphs should be short.
3. The body of the letter should have three parts: (1) statement of purpose, which is specified in the first paragraph; (2) the business of the letter, which could include answering a letter, asking a question, or giving information (this will be in one or more paragraphs and/or sections); and (3) summarization and closing, which will be in the last paragraph.
4. Bulleted or numbered lists should be used to break up long paragraphs and to highlight important information.
5. The letter should be written in the first-person format.
6. It should be respectful and written in a positive tone.
7. The active voice should be used when possible.
8. A letter should not be written when you are angry or emotional — if you must write, hold the letter for a day or two before mailing. This delay will give you time to cool down, and you can then make your decision to mail the letter in a more stable frame of mind.
9. The letter should be proofed by another person or two before mailing to make sure that you have not overlooked spelling or grammatical mistakes.
10. A letter becomes a "living testimony" to one's professionalism (or lack thereof). Unlike some things in the workplace, once a letter is written and transmitted it is written in concrete.

Introduction to Memorandums

While letters are used for formal and external correspondence, memorandums (memos) have long had a vital role in conducting day-to-day operations in business and industry for internal correspondence. Although someday the memo may be replaced by e-mail, it will be with us for at least another decade. Given the important role that the memo plays in commerce and the prediction that the use of the memo may be with us for another decade, it is instructive to review some of the uses and rules associated with the development of a memo. Due to the lack of technical writing courses and the wide use of the memo, many professionals do not fully understand that like any other technical document, there is a right time to use a memo and a wrong time.

First, the memo is an internal communications document, which means that you should not use it to communicate with your customers, vendors, or another contact outside your company. Because the memo is an informal type of communication, it is effective in communicating with your supervisors, employees, and other company colleagues and contacts without all of the time and trimmings associated with the traditional letter. With the preceding in mind, this chapter provides information and recommendations that can help technical professionals create quality memos that will be read.

When to Use the Memo

Generally there are at least 15 purposes for a memo (Maki and Schilling 1987, 156–157):

1. To confirm telephone conversations or previous correspondence.
2. To explain circumstances leading up to the present correspondence.
3. To clarify a misunderstanding or point of view.
4. To acknowledge your reader's position or point of view.
5. To argue against a particular point of view.
6. To reaffirm a commitment.
7. To flag a reader's attention or interest.
8. To describe an existing condition.
9. To summarize information.
10. To explain a problem.
11. To alert readers to an impending problem.
12. To announce a change.
13. To express personal response, such as congratulations, compliment, thanks, disappointment, or sympathy.
14. To boost morale.
15. To create an image of yourself and/or [sic] your organization.

Memo Format

Some companies have memo stationery or a standard memo format that should be followed, while other companies have no style or format rules associated with development of a memo. Whether your company uses special memo stationery or if you print your memo on blank paper, there are some common rules related to the format that should be used for memos:

- The word "Memorandum" should be at the top of the document so that the reader can readily identify the purpose of the document.
- The next line is for the date; unless your company uses a military format, it is best to write the date in a standard format, such as "March 23, 1995," which is better than "3/23/95."
- The next line of information must be the designation "To:" and should specify the names of all intended recipients. If the list is more than five individuals, it is appropriate to use group designations such as "Lead Engineers," "Quality Assurance Specialists," "Supervisors," etc. Also, as illustrated in Figure 7.C, to ensure that each name is readily identified, it is best if each one appears on a separate line alphabetically, or in order of hierarchy of authority. The individuals designated on the "To:" line are those who are directly affected by the information in the memo.
- The next line is for the "From:" and should be the name and title of the individual sending the memo.
- The following line is for the "Subject:" and should be a short descriptive title or subject. This line should not be cute or evasive — it should immediately inform the recipient why he or she should take time to read your memo.
- If the memo is to be copied to one or more individuals, then a "Copy:" must follow the "To:'" line. Each name should appear on a separate line similar to the format used for the "To:" line. The "copy" designation is for those individuals who may not be directly affected by the information provided in the memo but have a "need to know" requirement. No action is required by individuals listed on the "Copy" line.
- In rare instances, you may want to convey that the information or subject of your memo is of an "urgent" nature. Adding a "Priority: Urgent" line as the last header will generally ensure that the recipient immediately reads your memo. Use the "Priority: Urgent" line only in emergencies; if you use it with every memo it will be tantamount to the "boy who cried wolf" story (see Figure 7.C).
- As reflected in Figure 7.C, another line is used to separate the header information from the memo message.
- Ideally, a memo should be limited to one and generally no more than three pages and should contain important information.
- Follow all grammar and spelling rules when writing memos; they should not be written in shorthand format.
- Use short paragraphs.
- Use single space between lines and double space between paragraphs.
- Like other writing formats, use bullets or numbers to highlight important information.

Figure 7.C Memo Format

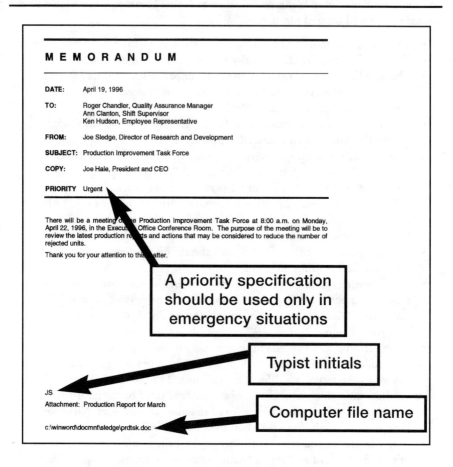

- Avoid using the passive voice.
- Like other writing formats, do not write a memo when you are emotionally upset or angry. If you feel that the memo must be written when you are angry, at least let it sit on your desk overnight before you send it to the addressee.
- In addition to the banner information and the memo body, there are four other elements that can be included at the end of the memo:

    — *Typist and sender initials* — these used to be a requirement for all work-related correspondence. Today, the use of the typist's and sender's initials is determined more by personal style as a larger percentage of professionals are word-processing their own memos. The use of initials can be helpful, however, when the memo is word processed by a secretary. The initials will help to quickly identify who word processed the memo in situations where a correction needs to be made, or the same memo can be used to send information to others where only name changes are required.

— *Enclosures* — this specifies all enclosures (enclosed in the envelope or package but not "attached") that will be transmitted with the memo. Enclosures are usually reserved for larger documents or items that constitute a separate entity, such as a book or report. Conventional practice has been to simply specify the number of enclosures so the reader will know that all intended items have been received. A safer practice, however, is to list by name each of the enclosed items so that there will be no misunderstandings about what has actually been sent. Such a practice is extremely critical when sending legal documents or verification of transmission is required.

— *Attachments* — this specifies all attachments that will be transmitted with the memo. Attachments are usually reserved for smaller documents or reports that can be attached by staple or paper clips. Often the memo is meant to be kept with the respective document.

— *Computer file name and directory name* — these are being used more as the computer is handled by almost every company employee. Including the computer file name and directory name will save a large amount of time in finding the file for use later.

How to Write Memos That Will Get Read

For memos of transmittal, commitment, confirmation, and other short topics, the body should simply include an introductory sentence, topic sentence or sentences, and closing sentences. For more complicated topics or situations where a longer memo is required, extreme care must be taken regarding content. *Mastering Memos*, a video from the editors of *Communications Briefings*, states that "most memos should never be written" (*Mastering Memos* 1992). The point is, that most of the information in memos in the workplace is meaningless or could have been more efficiently communicated in person. Because of the "worthless memo syndrome," the technical professional must first make *sure* that the memo is needed to convey important and meaningful information before beginning a draft. Once it has been determined that the memo is in the best format, the next logical step is to plan the memo in terms of who, what, when, and why questions:

- *Who* — must receive the memo?
- *What* — is the main subject or topic?
- *When* — did or will the event take place and when must people be notified?
- *Why* — is the subject or topic important?

Answering these four questions will help you collect all the necessary information to begin writing your memo. The actual writing process can be streamlined by using a method called *SOPPADA*. Although the origin of *SOPPADA* is not known, the method and its benefits are introduced in the *Mastering Memos* (1992) video. *SOPPADA* is actually an acronym where each letter stands for a specific element that must be included to develop a readable and informative memo. The *SOPPADA* translation is as follows:

Letter	Memo Element
S	Subject
O	Objective(s)
P	Present situation
P	Proposal
A	Advantages
D	Disadvantages
A	Action needed

Each of the *SOPPADA* elements is described as follows (see Figures 7.D.1 and 7.D.2 for a graphic representation of a *SOPPADA* memo):

- *Subject*— will be specified in the banner of the memo as described in the section "Memo Format." Essentially, the subject must be succinct and descriptive, giving the reader a reason to read it. Also, the subject descriptor must be directly correlated to the subject that is actually in the memo. A good practice to ensure that the subject has been properly described is to review the subject line after you have finished writing your memo to make sure there is a direct correlation between the memo information and the subject line.
- *Objective* — is presented in the very first paragraph to specify why you are writing the memo. It must be short and to the point so the reader will know exactly why the memo was transmitted and why he or she must read it.
- *Present situation* — is full details and specifications on the present situation and why a change is needed or an action must be taken. It is extremely important to be to the point and provide the reader with all related critical information.
- *Proposal* — is a detailed explanation of the proposal that you have formulated to address a resolution to the information or problem provided in the present situation.
- *Advantage(s)* — is a full description of all of the advantages that will occur as a direct result of your proposal. The advantages should be short and to the point.
- *Disadvantage(s)* — is a full description of all of the disadvantages that

could occur as a direct result of implementing your proposal. It is human nature for most people to avoid providing information on the disadvantages of a proposal. Providing a full description of the disadvantages, however, will demonstrate to the reader that you are forthright and "laying all the cards on the table," which has a disarming influence. Also, providing all the disadvantages up front may very well be your salvation if something goes wrong in the implementation of your proposal; you can always say that you were up front about it and that everyone knew the disadvantages that could result from implementation.

• *Action needed* — will inform the readers about what is required or asked of them. The action needed should be a separate paragraph and be clearly specified. Also, it is good practice to specify a deadline for the action needed. In situations where you are sending the memo to a superior, it is recommended that you follow normal etiquette and use such phrases as "please respond by" or "it would be helpful" and other phrases that are in line with traditional superior–subordinate etiquette in the workplace.

Figure 7.D.1 Sample Memo (Page 1 of 2)
Using the SOPPADA Formula

 **AERIAL DATA
REDUCTION
ASSOCIATES, INC.**

Photogrammetric Engineering • Geographic Information Systems • Geodetic
Control Surveys • GPS Surveys • CAD Hardware, Software and Support •
Photo Interpretation • Terrestrial Photogrammetry • Aerial Photography

INTEROFFICE MEMO

Date: April 12, 1996

To: Geodetic Services Field staff

From: John V. Hamilton, Director of Operations

Subject: Revised log sheets

Please note the attached revised *GPS Survey Data Log Sheet*. This form has
been revised to alleviate some of the data logging problems experienced in the
field by survey crews.

The current GPS log sheet has been shown to be somewhat difficult to complete
while in the field due to extreme time constraints during fast static GPS observa-
tions. The flow through the form does not match the typical flow of information as
presented and collected by the data receiver, and causes unnecessary delays in
"finding the right box" to log the next sequential data item.

As you will note, the requested information to be logged has been rearranged in
a more sequential fashion that reflects the actual sequencing in the field.

Basic flow through the form is left to right, top to bottom. The sketch area is now
moved to the back of the form to allow a more detailed sketch for later station
recovery. Also, two new items have been added for logging:

    1. satellites in use at time of observation

    2. initial ellipsoid height observations

The new form should allow more rapid completion of required data logging while
freeing up some session observation time to concentrate on the actual observation procedure parameters. Further, office data processing will flow more easily
due to the revised data groupings that require computer operator entry.

The primary drawback to the new forms is that of all new procedures: a change
in procedure must be accommodated, and the sequence flow of the data logging
must be somewhat practiced to fully realize the anticipated time savings in the
field. There will most likely be some station observation sessions with missing

Subject	
Objective	
Present situation	
Proposal	
Advantage(s)	
Disadvantage(s)	

RESPOND TO:
PA: 280 Kappa Drive, Pittsburgh, PA 15238   412/967-9577   Fax 412/967-9564

WITH OFFICES IN:
NJ: 9285 Commerce Highway, PO Box 557, Pennsauken, NJ 08110   800/257-7960   Fax 609/486-7778
PA: 280 Kappa Drive, Pittsburgh, PA 15238   412/967-9577   Fax 412/967-9564
CO: 1155 Kelly Johnson Blvd., Suite 110, Colorado Springs, CO 80920   719/590-9989   Fax 719/590-9070
AL: 600 Boulevard South, Suite 104, Huntsville, AL 35802   205/883-3508   Fax 205/883-3507
CT: PO Box 290955, Wethersfield, CT 06129   800/257-7960
FL: 2506 Maplewood Drive, West Palm Beach, FL 33415 800/468-4670
RI: 33 Alexander Avenue, East Providence, RI 20914   401/434-0134

Figure 7.D.2 Sample Memo (Page 2 of 2)
Using the SOPPADA Formula

data, particularly concerning the two newly requested items (satellites in use and initial ellipsoid heights). This should not be a major cause of concern, however; just be aware that they will become required log items once the new system is fully in place.

ACTION REQUIRED: remove all previous edition blank log sheets from your field log books and replace with this new revision, ADRGS.LOG rev 0496–02 on the next **new project** started. It is expected that some transition time will become necessary to get the feel for using the new log sheets, so at this point pay careful attention to all data logging. Remember the data flow is now more uniformly left to right, top to bottom.

The new log sheets should be in use fully by June 1, 1996, unless a project is already in progress using the old format, for which you should continue to use the old log sheets until the project is complete.

JVH

Attachments: 1) revised log sheet ADRGS.LOG rev 0496–02

REVISED GPS LOGS   2

Disadvantage(s) (cont'd)

Action needed

CHAPTER
8

SCIENTIFIC AND
TECHNICAL REPORTS

*"Experts agree on the key to an effective report. It is important to
know the objectives of your report, who your readers are and what
they want to gain from your report" (Herbert 1994, 34).*

Sarah Herbert
Progress Reports
Engineering

Introduction

Scientific and technical reports (hereafter referred to as reports) play an extremely
important role in today's workplace. Equally important is the undisputed fact that the
report is the most important document that technical professionals will be required
to write during their careers. With the report, supervisors, corporate executives,
boards of directors, clients, and even the public are informed and kept up to date about
progress on projects and even important findings on research and development projects.
The report usually is developed in an industry-standard format. Today, there are perhaps
20 or more different types of reports written in the engineering, scientific, and technical
fields. As observed by Thielsch (1996) one of the world's leading failure analysis experts
in the engineering field, typical groups of reports include "(1) field engineering
evaluation reports, (2) failure reports, (3) equipment condition assessment reports,
(4) equipment safety reports (risk management reports), (5) engineering project cost
reports (quantum reports), (6) inspection reports with or without engineering
interpretations, (7) compliance reports covering equipment compliance standards,
specifications, governmental agency requirements."

Regardless of the specific type of report, information in all reports is essentially
presented in a consistent format where the readers can easily identify the major
elements of the report: (1) Cover; (2) Title page; (3) Abstract; (4) Table of contents;
(5) List of figures and tables (if applicable); (6) Introduction; (7) Report core, which
includes methodologies, assumptions, and procedures; (8) Results and discussion;
(9) Conclusion(s); (10) Recommendations; (11) References; and (12) Support
material, which includes appendices, glossary, and index.

The remaining portions of this chapter provide information, recommendations, and industry samples for the two major groupings of reports, that the technical professional will be required to develop: (1) scientific reports; and (2) technical reports. The major differences between a scientific report and a technical report is that a scientific report will generally focus on the results of a study or experiment. As is described here, a scientific report will include an expanded discussion on "analysis of the data" in the sections on "results and discussion" and "recommendations."

Again, because there are so many types of reports, this chapter focuses on those elements that in terms of practice and industry standards should be the minimum. Additional elements and information are acceptable as long as they provide needed information to the intended audience. Thus, as in all forms of technical writing, the technical professional must always keep the target audience in mind when developing reports.

Industry Standards

There are two excellent industry standards on the market that deal with the development and presentation of scientific and technical reports. The first standard is titled *Documentation — Presentation of Scientific and Technical Reports* (ISO 5966) (ISO 1982)(see Figure 8.A) published by the International Organization for Standardization (ISO). The ISO is a worldwide federation of national standard organizations from almost 200 countries. Established in 1947, ISO has a current mission "to promote the development of standardization and related activities in the world with a view to facilitating the international exchange of goods and services, and to developing cooperation in the spheres of intellectual, scientific, technological and economic activity" (ISO 1996, 1). The development of the standards is done by some "2,700 technical committees, subcommittees and workgroups" (ISO 1996, 4).

This standard *Documentation — Presentation of Scientific and Technical Reports,* also referred to as *ISO 5966,* includes 22 pages of information on the development and layout of reports. Included is information on the numbering of the report, as well as descriptions of all major elements (e.g., title page, table of contents, core report, conclusions and recommendations, etc.). Useful information is also provided on how to work with graphics (e.g., graphs, illustrations, photographs, etc.) and brief, yet helpful, specifications on the production of the report (e.g., reproduction, binding, etc.). Although of value to all types of organizations, ISO 5966 is of particular value to those doing business on an international level.

Another relevant ISO standard is *Information and Documentation — Bibliographic References—Electronic Document or Parts Thereof.* Although this standard, also known as *ISO 690-2,* is currently in draft form, it is extremely useful in obtaining an acceptable reference format of cited material from electronic documents. As many of the style manuals, including the *Chicago Manual of Style* cited in Chapter 1, work closely with the ISO, this standard will eventually be used by most publishing organizations.

Figure 8.A ISO 5966

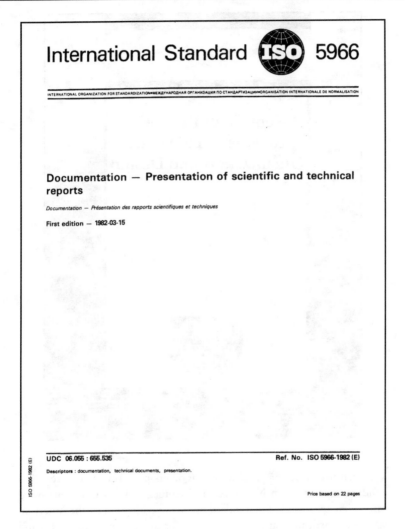

International Standard **ISO** 5966

INTERNATIONAL ORGANIZATION FOR STANDARDIZATION●МЕЖДУНАРОДНАЯ ОРГАНИЗАЦИЯ ПО СТАНДАРТИЗАЦИИ●ORGANISATION INTERNATIONALE DE NORMALISATION

Documentation — Presentation of scientific and technical reports

Documentation — Présentation des rapports scientifiques et techniques

First edition — 1982-03-15

UDC 06.055 : 655.535

Descriptors : documentation, technical documents, presentation.

Ref. No. ISO 5966-1982 (E)

Price based on 22 pages

ISO 5966-1982 (E)

The second industry standard related to development, layout, and presentation of reports is titled *Scientific and Technical Reports — Elements, Organization, and Design* (see Figure 8.B). This standard was developed by the National Information Standards Organization (NISO) and approved by the American National Standards Institute (ANSI). It has been published under the National Information Standards Series and is best described by the publishers as follows:

> NISO standards are developed by the Standards Committees of the National Information Standards Organization. The development process

Figure 8.B ANSI/NISO Z39.18-1995

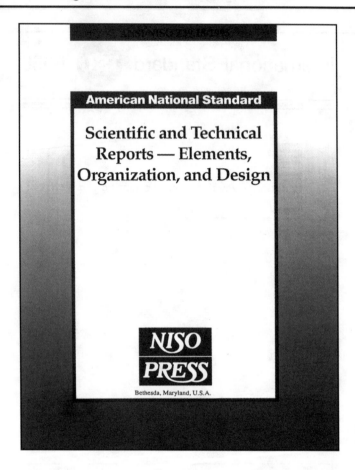

is a strenuous one that includes a rigorous peer review of proposed standards open to each NISO Voting Member and any other interested party. First approval of the standard involves verification by the American Standards Institute that its requirements for due process, consensus, and other approval criteria have been met by the NISO. Once verified and approved, NISO Standards also become American National Standards.

The use of an ANSI/NISO Standard is voluntary. That is, the existence of this NISO standard does not preclude anyone, whether or not that person has adopted the NISO standard, from manufacturing, marketing, purchasing, or using products, processes, or procedures that do not conform to the NISO standard. However, the use of standards (those developed by NISO as well as other standards-developing organizations) has proven to be in the best interest of any industry wishing to increase its effectiveness and efficiency in the areas of product development, manufacturing, and marketing and, therefore, such use is encouraged by ANSI, NISO, and all other standards-developing organizations.

Every NISO standard is reviewed at least once every five years to confirm that NISO Standard remains viable and useful in its current environment. (NISO 1995, inside cover)

The *Scientific and Technical Reports — Elements, Organization, and Design* standard *(ANSI/NISO Z39.18-1995)* is an exceptional document that details all major report elements (e.g., title page, table of contents, core report, conclusions and recommendations, etc.). It includes useful information on how to work with graphics (e.g., graphs, illustrations, photographs, etc.) and an annotated bibliography of major technical writing books, style manuals, and other references.

While both standards, *ISO 5966* and *ANSI/NISO Z39.18-1995,* specify similar report elements, the latter is more descriptive and thus more functional. However, both are excellent standards and can be used to ensure that the report is developed in a logical and consistent format to communicate with the intended audience. The remaining portion of this chapter was developed using the recommended elements specified in the two discussed standards.

Report Elements

This section provides information and recommendations related to the following 13 elements that are normally associated with scientific and technical reports:

(1) Cover;
(2) Title page;
(3) Abstract;
(4) Table of contents;
(5) List of figures and tables (if applicable);
(6) Introduction;
(7) Report core which includes methodologies, assumptions, and procedures;
(8) Results and discussion;
(9) Conclusion(s);
(10) Recommendations;
(11) References;
(12) Support material, which includes appendices, glossary, and index; and
(13) Binding.

As reflected in Table 8.A, some of these elements are present in both scientific and technical reports, while other elements are included only in either the scientific or the technical report.

Table 8.A Required Elements for Scientific and Technical Reports

Report Element	Include in a Scientific Report	Include in a Technical Report
Cover	Yes	Yes
Title page	Yes	Yes
Abstract or executive summary	Yes	Yes
Table of contents	Yes — for reports over 10 pages	Yes — for reports over 10 pages
List of figures and tables	Yes — for reports over 10 pages	Yes — for reports over 10 pages
Introduction	Yes	Yes
Methodologies, assumptions, and procedures	Yes — this section will be more detailed for scientific reports than for technical reports	Yes
Results and discussion	Yes	Yes
Conclusion(s)	Yes	Yes
Recommendations	Yes	No — unless recommendations are required to rectify a problem or problems identified through the respective investigation and analysis process
References	Yes — if external sources and data are cited in the report	Yes — if external sources and data are cited in the report
Support material	Only if applicable	Only if applicable
Binding	Yes	Yes

In most instances, the report will be developed in the same format, which includes the following 12 elements:

1. *Cover*

   Some organizations see the cover as just window dressing with no value to the report. Such an assumption could not be further from the truth. As discussed in other chapters, the appearance of the document can be just as important as the information provided within. Beyond simply protecting the document, a properly developed and attractive cover will help attract the reader's attention. Also, the cover helps protect the report from damage that can occur through mailing and handling.

   As a minimum there are four elements that should be included on the cover:
   a. Descriptive title of the report.
   b. Name of the report author(s), principal investigator(s), if applicable, and editor, if applicable.

Figure 8.C Sample Report Cover From the U.S. Nuclear Regulatory Commission

NUREG/CR–6438
BMI–2188

The Effect of Cyclic and Dynamic Loads on Carbon Steel Pipe

Unique report number
issued by the NRC

Prepared by
D. L. Rudland, P. M. Scott, G. M. Wilkowski

Battelle

Prepared for
U.S. Nuclear Regulatory Commission

---

    c. Publication number — if an organization's format is not required, consider using the standard Documentation — Presentation of Scientific and Technical Reports (ISO10444).

    d. Publication date (optional, but recommended).

A sample report cover is provided in Figure 8.C.

2. Title Page

The first page of the report is generally called the title page and contains a descriptive title of the report, author's name, and company name and address (see Figure 8.D). Additional information, such as who the report was developed for, can be included on the title page if appropriate (e.g., customer's name, company and address, the board of directors, etc.). Although there may be some duplication of information from the cover

Figure 8.D Sample Title Page From
the U.S. Nuclear Regulatory Commission

NUREG/CR–6438
BMI–2188

The Effect of Cyclic and Dynamic Loads on Carbon Steel Pipe

Manuscript Completed: February 1996
Date Published: February 1996

Prepared by
D. L. Rudland, P. M. Scott, G. M. Wilkowski

Battelle
505 King Avenue
Columbus, OH 43201–2693

M. Mayfield, NRC Project Manager

Prepared for
Division of Engineering Technology
Office of Nuclear Regulatory Research
U.S. Nuclear Regulatory Commission
Washington, DC 20555–0001
NRC Job Code D2060

(used for appearance and to protect the document), all of the specified information must be included on the title page.

3. *Abstract*

The abstract should be a short description of the report including overview, findings, and recommendations when applicable (see Figure 8.E.1). The abstract must be a separate page of no more than 500 words, with 200 being the preferred number. If the abstract is used for a federal project, then it should be no more than 200 words as required in Standard Form 298 (see Figure 8.E.2). As the abstract may be used by company libraries and information services for indexing purposes, it must stand on its own merits to inform the readers of the purpose, findings, and recommendations of the report. Although permitted if required for explanation purposes, abstracts generally do not contain illustrations, figures, tables, or other

Figure 8.E.1 Sample Abstract From
the U.S. Nuclear Regulatory Commission

Abstract

ABSTRACT

This report presents the results of four 152-mm (6-inch) diameter, unpressurized, circumferential through-wall-cracked, dynamic pipe experiments fabricated from STS410 carbon steel pipe manufactured in Japan. For three of these experiments, the through-wall crack was in the base metal. The displacement histories applied to these experiments were a quasi-static monotonic, dynamic monotonic, and dynamic, cyclic (R = -1) history. The through-wall crack for the third experiment was in a tungsten-inert-gas weld, fabricated in Japan, joining two lengths of STS410 pipe. The displacement history for this experiment was the same history applied to the dynamic, cyclic base metal experiment. The test temperature for each experiment was 300 C (572 F).

The objective of these experiments was to compare a Japanese carbon steel pipe material with United States pipe material, to ascertain whether this Japanese steel was as sensitive to dynamic and cyclic effects as United States carbon steel pipe.

In support of these pipe experiments, quasi-static and dynamic, tensile and fracture toughness tests were conducted. An analysis effort was performed that involved comparing experimental crack initiation and maximum moments with predictions based on available fracture prediction models, and calculating J-R curves for the pipe experiments using the η-factor method.

graphics. It should be single-spaced in paragraph form and follow all other technical writing conventions and rules. Also, some organizations use the abstract as an executive summary.

4. *Table of Contents and List of Figures*
Immediately following the abstract will be the table of contents, which should be included in reports that are over 10 pages. As discussed in Chapter 4, today's high-end word-processing programs can automate the process of developing an attractive table of contents as reflected in Figure 4.D.1. The attractive feature of the computer-generated table of contents is that it can be easily updated as material is deleted or added to the report. Following the table of contents will be a list of figures and tables, which is generally used when there are more than five tables or figures for reader reference (see Figures 8.F.1 and 8.F.2)

Figure 8.E.2 Sample Completed Standard Form 298 From the National Aeronautics and Space Administration

REPORT DOCUMENTATION PAGE		Form Approved OMB No. 0704-0188
Public reporting burden for this collection of information is estimated to average 1 hour per response, including the time for reviewing instructions, searching existing data sources, gathering and maintaining the data needed, and completing and reviewing the collection of information. Send comments regarding this burden estimate or any other aspect of this collection of information, including suggestions for reducing this burden, to Washington Headquarters Services, Directorate for Information Operations and Reports, 1215 Jefferson Davis Highway, Suite 1204, Arlington, VA 22202-4302, and to the Office of Management and Budget, Paperwork Reduction Project (0704-0188), Washington, DC 20603.		

1. AGENCY USE ONLY (Leave blank)	2. REPORT DATE January 1996	3. REPORT TYPE AND DATES COVERED Technical Paper

4. TITLE AND SUBTITLE	5. FUNDING NUMBERS
Chaos in a Fractional Order Chua System	WU–505–62–50
6. AUTHOR(S) Tom T. Hartley, Carl F. Lorenzo, and Helen Killory Qammar	

7. PERFORMING ORGANIZATION NAME(S) AND ADDRESS(ES)	8. PERFORMING ORGANIZATION REPORT NUMBER
National Aeronautics and Space Administration Lewis Research Center Cleveland, Ohio 44135–3191	E–9532

9. SPONSORING/MONITORING AGENCY NAME(S) AND ADDRESS(ES)	10. SPONSORING/MONITORING AGENCY REPORT NUMBER
National Aeronautics and Space Administration Washington, D.C. 20546–0001	NASA TP–3543

11. SUPPLEMENTARY NOTES
Tom T. Hartley, Department of Electrical Engineering, and Helen Killory Qammar, Department of Chemical Engineering, University of Akron, Akron, Ohio 44325 (work funded by NASA Grant NAG3–1491); Carl F. Lorenzo, NASA Lewis Research Center. Responsible person, Carl F. Lorenzo, organization code 2500, (216) 433–3733.

12a. DISTRIBUTION/AVAILABILITY STATEMENT	12b. DISTRIBUTION CODE
Unclassified - Unlimited Subject Categories 66 and 31 This publication is available from the NASA Center for Aerospace Information, (301) 621–0390.	

13. ABSTRACT (Maximum 200 words)
This report studies the effects of fractional dynamics in chaotic systems. In particular, Chua's system is modified to include fractional order elements. Varying the total system order incrementally from 2.6 to 3.7 demonstrates that systems of "order" less than three can exhibit chaos as well as other nonlinear behavior. This effectively forces a clarification of the definition of order which can no longer be considered only by the total number of differentiations or by the highest power of the Laplace variable.

14. SUBJECT TERMS			15. NUMBER OF PAGES 24
Chaos; Dynamics; Nonlinear; Fractional calculus; Fractional order; Systems; Chua; Nonlinear dynamics; Fractional dynamics			16. PRICE CODE A03

17. SECURITY CLASSIFICATION OF REPORT Unclassified	18. SECURITY CLASSIFICATION OF THIS PAGE Unclassified	19. SECURITY CLASSIFICATION OF ABSTRACT Unclassified	20. LIMITATION OF ABSTRACT

NSN 7540-01-280-5500	Standard Form 298 (Rev. 2-89) Prescribed by ANSI Std. Z39-18 298-102

5. Introduction

The introduction must include a statement of purpose for the report, along with a statement on the scope. Also, if the report's target audience is nontechnical or mixed, then a good practice would be to include definitions of technical terms and abbreviations used in the body of the report.

6. Methodologies, Assumptions, and Procedures

This section includes a detailed specification of methodologies, procedures, and assumptions used to conduct the study or experiment. Be sure to keep in mind your target audience when writing the procedures

Figure 8.F.1 Sample List of Figures From the U.S. Nuclear Regulatory Commission

section. Finally, include only information pertinent to the readers — remember that time is a valuable resource and you do not want to lose their interest before they read the other, more important, sections of your report.

7. Results and Discussion

The results and discussion section, sometimes simply called "discussion," consists of the findings of project activities, which could be research based or simply a report of other relevant findings. The results must be clear, concise, and logically ordered so as not to lose the readers. The results section could also include tables, graphs, and other graphics to help support the text information (see Figures 8.G.1 through 8.G.5 for a partial sample).

Figure 8.F.2 Sample List of Tables From the U.S. Nuclear Regulatory Commission

8. Conclusion(s)

The conclusion(s) section, sometimes called "summary and conclusions," basically provides a summary of the major answers and findings described in the analysis section. Again, the information provided in this section should be concise and presented in a logical order (see Figures 8.H.1 and 8.H.2).

9. Recommendation(s)

This sections includes recommendations to help rectify problems or findings specified in the analysis and conclusion sections. The writer should provide recommendations only if empowered to do so by the proper authorities. Additionally, the recommendation(s) section could include a specification of additional studies that could be undertaken to improve the understanding of the problems presented in the report. Additionally, only recommendations pertinent to the study or project should be

Figure 8.G.1 Sample Portions (Page 1 of 5) of a Report From the National Aeronautics and Space Administration

INTRODUCTION

Within the international community of aeronautical–transportation specialists there are important issues that require careful study before commitment to a high-speed air transport concept can be made. These issues include speed, range, seat capacity, propulsion concepts, and the effect on the environment, as well as materials, structures, stability and control, safety, and operational concerns.

With these and other related issues in mind, the European Symposium on Future Supersonic–Hypersonic Transportation Systems was held in Strasbourg, France in November, 1989. This symposium was organized by L'Académie Nationale de l'Air et del l'Espace (A.N.A.E., Fr.) and Deutsche Gesellschaft für Luft und Raumfart (D.G.L.R., Ge.) with the participation of the Royal Aeronautical Society (R.Ae.S., U.K.). Technical presentations were given by members of these societies and by participants from other European nations as well as by American and Japanese specialists.

More than 40 presentations were given during the 2 1/2-day symposium including 2 papers representing the flight experience and proposed activities of the NASA Dryden Flight Research Facility. These two presentations, "The Need for a Hypersonic Demonstrator" and "NACA–NASA Supersonic Flight Research" were delivered by Mr. Theodore G. Ayers, then Deputy Director of the Dryden Flight Research Facility (DFRF). The two DFRF presentations are represented in print in reference 1, which is an edited compilation of the proceedings of the previously mentioned symposium.

In reference 1, the NASA DFRF presentations are printed without references which somewhat diminishes their usefulness and deprives the reader–researcher of interesting historical and background information. In addition, the Supersonic paper is printed from a transcription of the tape recording taken during the symposium. Thus the rendition given in reference 1 is in the first person and in a predominantly oral format. For the reasons cited, and because a wider distribution of the Supersonic paper is considered appropriate, this expanded document has been prepared.

The NASA DFRF evolved through several agency reorganizations beginning in 1946. The original organization was the Muroc Flight Test Unit of the National Advisory Committee for Aeronautics (NACA).[*] The Muroc Flight Test Unit began as a small group of engineers and technicians from the NACA Langley Aeronautical Laboratory who were transferred to the Muroc Army Air Field in California's Mojave Desert in 1946.[2] The NACA Muroc Flight Test Unit was formed to provide technical guidance for testing the Army/Bell XS-1 rocket aircraft (later known as the X-1). The X-1 was to become the first manned airplane to exceed the speed of sound on October 14, 1947.

This report begins with the early X-1 series of research aircraft, and provides selected examples of supersonic flight research from the next four decades. This document begins with the transonic-supersonic demonstration and evaluation of the adjustable horizontal stabilizer on the X-1 airplane. Twelve other examples of supersonic flight research follow, which include aircraft efficiency, stability and control, structural loads, model-to-flight correlation, parameter estimation, and more recent developments leading to the integration of flight and propulsion controls and digital fly-by-wire technology.

[*]Later names for the organization were, in order, High-Speed Flight Research Station, High-Speed Flight Station, Flight Research Center, Dryden Flight Research Center, and the Dryden Flight Research Facility. On March 1, 1994 the organization was reestablished as the Dryden Flight Research Center.

specified. The recommendations should be listed in numerical order similar to the way the findings were specified in Figures 8.H.1 and 8.H.2.

10. References

Full bibliographic references must be provided for external sources, data, and other information cited in the report, including text, data, tables, figures, and other graphics. Reference specification must begin on a new page with the heading "Reference" (see Figure 8.I). All references must follow the respective style manual mandated by the company, organization, or discipline. If no style is mandated or recommended, then the latest edition of the *Chicago Manual of Style* should be used.

11. Support Material

The support material and references include all additional or detailed information the reader *may* refer to related to the subject matter. The

Figure 8.G.2 Sample Portions (Page 2 of 5) of a Report From the National Aeronautics and Space Administration

As the title of this report indicates, the research described herein was either obtained at supersonic speeds or enabled subsequent aircraft to penetrate or traverse the supersonic region. Consequently, this report does not address significant flight research accomplished at subsonic speeds or at hypersonic speeds.*

While the types of aircraft in this report vary widely, and a broad range of aeronautical disciplines are discussed, many NASA-DFRF aircraft–programs are not included. The 13 examples presented are categorized into 3 specific stages of aeronautical flight research and were chosen to show that flight research has evolved through several stages since the post World War 2 years. The three stages which represent the DFRF supersonic flight research experience are

Stage 1   Barriers to Supersonic Flight

Stage 2   Correlation–Integration of Ground Facility Data and Flight Data

Stage 3   Integration of Disciplines

The relevance of each stage of aeronautical research and the 13 examples will be highlighted in a summary following each stage in the Discussion and in the Concluding Remarks section. The table on page 58 lists the 13 selected examples of supersonic flight research in relation to the 3 research stages. In the meantime, the authors hope that the aircraft types, the aeronautical disciplines, and the solutions to problems presented herein will reveal the significance of exposing the problems of existing airplanes and the advance design concepts of future aircraft to the realities of the flight environment. Because, to quote Dr. Hugh L. Dryden, the purpose of full-scale flight research is "... to separate the real from the imagined ... to make known the overlooked and the unexpected problems."

DISCUSSION

Stage 1: Barriers to Supersonic Flight

Initially the most obvious of the barriers to achieving supersonic flight were the rapid changes in control effectiveness and the sudden onset of wave drag near Mach 1. Though the first manned flight to exceed sonic speed did not by itself solve the problems associated with these barriers, it demonstrated that their ultimate solution was likely. As indicated in the introduction, that first flight to exceed the speed of sound was made by the X-1-1 on October 14, 1947.[2]

The first two X-1 airplanes had the same fuselage and planform configurations, but the thicknesses of the wings and horizontal stabilizers were different. Before the design was finalized there were two opposing philosophies regarding wing thickness. One group wanted a maximum thickness of 12 percent of chord. That way, detailed pressure data could be obtained for supercritical flow at an aircraft Mach number significantly below one and such a wing would be stronger than a thin wing. The other faction wanted a maximum thickness of 5 percent of chord so as to penetrate deeply into the supercritical region

*The definition of hypersonic, as applied to this report, is $M \geq 5.0$. Thus supersonic will refer to Mach numbers from 1.0 to less than 5.0.

support material could include appendices with expanded tables and graphs or even an expanded technical discussion for reports written for mixed audiences. In such an example, the longhand calculations of the data analysis could be included in appendices, where only answers or solutions were provided in the text of the report. Thus, where the solutions were presented, the readers would be told to refer to the appendices for the calculations used for the specific solution. Also, in some reports, support materials may be added when legal documentation is required.

12. Bindings

Rather than using staples, all reports should be professionally bound to enhance the presentation and preserve content. There are many different binding systems, ranging from comb binding to the more expensive glue systems. A more practical approach is to use a three-ring binder that

Figure 8.G.3 Sample Portions (Page 3 of 5) of a Report From the National Aeronautics and Space Administration

with lower shock strength (thus attenuating transonic nonlinearities); perhaps even permitting supersonic flight.[3] A compromise was reached which resulted in the following thicknesses.

Aircraft	t/c for wing, percent	t/c for horizontal stabilizer, percent
X-1-1	8	6
X-1-2	10	8

For a specific airplane, the thickness of the horizontal stabilizer was less than for the respective wing at the insistence of NACA advisors. This was so that the stabilizers would not experience transonic (shock-stall) problems simultaneously with the respective wing.

Not surprisingly, it was the X-1 airplane having the thinner wing that first flew supersonic. This airplane is shown in figure 1 along with a reproduction of the recording traces of the first "Mach jump." Note also the Mach diamond pattern in the rocket exhaust. The upper trace showing the jump is a history of impact pressure (q_c) i.e., the difference between the stagnation pressure sensed at the tip of the noseboom head and the static pressure sensed from flush orifices. The flush orifices were located several inches behind the tip of the noseboom head and the static pressure sensed is shown as the lower trace, labeled 5H. The time scale is indicated by the numbers beginning at 145 sec with time advancing to the right.

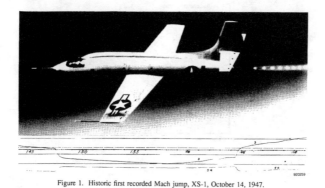

Figure 1. Historic first recorded Mach jump, XS-1, October 14, 1947.

will not only last longer than many other types of binding systems but will also lie flat when opened for ease of reading. Additionally, the binders can be purchased in a wide variety of colors and styles including the type with clear plastic sleeves on the covers and spines to insert custom covers (see Chapter 12).

Writing Scientific and Technical Reports

There are at least 16 recommendations that should be followed when writing a laboratory study or progress report, including:

1. Determine your target audience. As suggested in Chapter 2, if you have a mixed audience (technical and nontechnical), consider writing the report

Figure 8.G.4 Sample Portions (Page 4 of 5) of a Report From the National Aeronautics and Space Administration

The first jump was caused as the primary bow shock wave generated by the fuselage nose passed over the flush orifices at about 147 sec. As the airplane decelerated the bow shock passed over the flush orifices again in the opposite direction at about 164 sec. Though the recorded jump interval was about 17 sec, the airplane was actually slightly supersonic for 20.5 sec when an accounting was made for the effects of compressibility on the measured static pressure. The maximum Mach number reached during this flight was 1.06.

This airplane was to reach higher Mach numbers during later flights, though it never ventured far into the supersonic region because of its limited fuel capacity. Nevertheless, this airplane and its sister craft, the X-1-2, contributed significantly to all subsequent aircraft which performed either within or beyond the transonic speed region. A compilation of early research from the X-1-1 and the thicker winged X-1-2 is found in reference 4.

Figure 2 shows the relationship of the maximum Mach numbers obtainable for the early rocket-propelled research aircraft to the Mach capability of contemporary fighter aircraft over a period of about one decade.[2] The lead time for the research aircraft shown was approximately 5 years. This lead was established not only by the earliest X-1 airplanes but by other follow-on research aircraft. The lead in years demonstrated by these unique high-performance research aircraft illustrates how knowledge and concepts developed through such facilities provided the basis and confidence for increasing Mach capability of subsequent operational military aircraft.

Though these unique high-performance research aircraft extended the energy boundaries of achievable Mach number and altitude, more conventional aircraft are also important to conduct flight research.

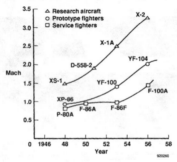

Figure 2. Leader-follower relationship between research aircraft, military fighter prototypes, and military fighters in service, reference 2.

for the nontechnical audience and provide more detailed technical data and information in an appendix.

2. Prior to writing the report, develop a detailed outline that specifies the information the readers need to know. The outline must be developed in a logical order.

3. Keep in mind that the purpose of your report is to inform, not to impress.

4. Double check all data and information for accuracy before writing. Be sure the data are reported accurately and have not been manipulated to support the writer's perspective. Sometimes the writers will struggle to make the data fit their perspective. Let the results of the data fall where they may because you will not be able to fool other professionals very long. If the data are manipulated, inevitably someone will stumble across the discrepancies; such discoveries have ruined the careers of otherwise competent professionals. Do not forget the old saying, "Figures don't lie

Figure 8.G.5 Sample Portions (Page 5 of 5) of a Report From the National Aeronautics and Space Administration

Modified, highly instrumented, operational aircraft are often used to fill in significant details to the technical fabric of aeronautical knowledge. It will become evident that both kinds of aircraft (unique specially built and modified-operational) are required to fulfill the purpose of flight research as defined earlier by Dr. H.L. Dryden, page 6.

The Adjustable "All-Movable" Stabilizer—It was apparent after World War 2 that to fly supersonically, airplanes would have to maintain control through regions characterized by rapid trim changes and diminished control effectiveness. Consequently the X-1 airplanes and the follow-on X-2[5] were provided with an in-flight, adjustable (all-movable) horizontal stabilizer, at the insistence of NACA, to compensate for the anticipated loss in elevator effectiveness. The earliest versions of the tiny X-1 airplanes did not have room for auxiliary power sources to operate a hydraulic system. Therefore, this pioneering transonic adjustable stabilizer had to be actuated by other means. The initial approach used a 24-volt battery-powered electric motor to drive a screw jack which changed the stabilizer incidence angle. Following a few low-speed glide flights it was decided that faster rates of change in incidence angle were needed. Consequently pneumatic motors were placed at each end of the screw jack, and these provided rates that were adequate for the transient conditions that would be encountered during powered flight. These pneumatic motors were driven by gaseous nitrogen.[6,7] Gaseous nitrogen at various pressure levels was also used to lower the landing gear, operate the flaps, deliver propellants to the rocket motor, operate the gyros, and pressurize the cockpit.

An example of data showing the benefits of the movable stabilizer is shown in figure 3. The left portion of figure 3 shows trim curves obtained at various horizontal stabilizer settings (i_t) for the X-1-2, corrected to a constant normal force coefficient of 0.3.[8] The data show that for stabilizer settings near 1° the airplane could be trimmed by the elevator (δ_e) through the speed range of Mach numbers above 1.0. However, if the stabilizer is set at 2°, there is insufficient up-elevator for control of the "tuck";* and at a setting of 0.5° there is too little down-elevator to overcome the nose-up trim change. The trim curve for the stabilizer, with $\delta_e = 0°$, is shown by the dashed line (where the ordinate scale applies).

The effectiveness of this control surface as a function of Mach number is shown in the right-hand portion of figure 3. These data together with other data, not shown, indicate that at Mach 1, the elevator was only 1/20 as effective as the stabilizer. Subsequently there resulted the general adoption of all-movable stabilizers on later high-speed airplanes ... first the X-2, the X-3, and the F-86 with all-movable horizontal stabilizers and later the X-15 research airplane having horizontal and vertical stabilizers all-movable. Virtually all supersonic fighter-interceptors flying four decades later use the all-movable stabilizer concept as well as larger cruise aircraft which probe even slightly into the transonic region. Current and future supersonic and hypersonic aircraft, manned or drone, also must rely on all-movable stabilizers; a concept which was first evaluated in transonic–supersonic flight in the late 1940's.

*Tuck as used here was the tendency for the nose of an airplane to rotate downward in pitch.

but liars figure," because others won't.

5. Include only necessary and important data in the body of the report. Remember that the appendices can be used for expanded data presentation if necessary.

6. Be thorough but concise.

7. Although not required, it is recommended that sections (introduction; methodologies, assumptions, and procedures; results and discussion; conclusion(s); and recommendations) and subsections be presented in an outline format. Under such a numbering system, whole numbers (1, 2, 3, etc.) reflect major topics such as "Introduction," "Methodologies, Assumptions, and Procedures," and "Results and Discussion," while the decimal side of whole numbers (1.1, 1.1.1, 2.1, etc.) reflects the subsections of each of the sections. This number format is used widely in almost all industries for all types of technical documents. Additionally,

Figure 8.H.1 Sample Summary and Conclusions (Page 1 of 2) of a Report From the U.S. Nuclear Regulatory Commission

---

Section 6   SUMMARY AND CONCLUSIONS

6.0 SUMMARY AND CONCLUSIONS

As part of this program, three dynamic and one quasi-static pipe fracture experiments were conducted on sections of 152-mm (6-inch) nominal diameter, STS410 carbon steel pipe manufactured in Japan. Three of the experiments involved through-wall cracks in the base metal of the STS410 carbon steel material. The fourth involved a through-wall-crack in a TIG weld joining two sections of the same STS410 carbon steel pipe material. All were loaded in four-point bending. The load histories for the four experiments were: (1) dynamic, monotonic with the crack in the base metal, (2) dynamic, cyclic with a stress ratio (R) of -1 and the crack in the base metal, (3) dynamic, cyclic with a stress ratio (R) of -1 with the crack in the center of a tungsten-inert-gas (TIG) weld, and (4) quasi-static, monotonic with the crack in the base metal. The test temperature for each experiment was 300 C (572 F). Each test specimen was unpressurized.

In addition to the pipe experiments, quasi-static and dynamic, tensile and fracture toughness properties for both the base metal and weld metal were evaluated. Also, a number of analyses were conducted for each pipe experiment. The moments at crack initiation were predicted using the GE/EPRI and LBB.ENG2 J-estimation schemes and the R6 Revision 3 Option 1 method and were compared with the experimental values. The maximum moments for each experiment were predicted using the Net-Section-Collapse, Dimensionless-Plastic-Zone-Parameter, GE/EPRI, LBB.ENG2, and R6 Revision 3 Option 1 methods and were compared with the maximum moments from the experiments. Finally, the J-R curves from the pipe experiments were calculated using the η-factor method and the load-displacement test record from each of the pipe experiments.

As a result of these efforts, the following conclusions can be drawn:

(1)   The STS410 carbon steel pipe material and the associated TIG weld studied in this program may be slightly susceptible to dynamic strain aging effects, although the extent of their susceptibility is probably less than some of the U.S. manufactured carbon steel base metal materials evaluated in the Degraded Piping, IPIRG-1, and Short Cracks in Piping and Piping Welds programs previously conducted at Battelle. The absence of any crack instabilities, i.e., crack jumps, in any of the pipe fracture experiments also shows that the pipe material tested was not highly sensitive to dynamic strain aging effects.

(2)   Based on a comparison of maximum loads for the three base metal experiments, the STS410 carbon steel pipe material appears to be less sensitive to cyclic loading effects than the A106 Grade B pipe material evaluated in Subtask 1.2 of the IPIRG-1 program. However, when one compares the J-R curves for the pipe experiments from the η-factor analysis, one sees a dramatic reduction in toughness due to cyclic loading, see Figure 4.6. This material seems to be more sensitive to cyclic loading after some amount of crack growth than at crack initiation. This may explain why the maximum load predictions for the three base metal experiments agree so closely. The maximum load may have been attained before significant crack growth occurred.

---

this numbering format is recommended in ISO 2145-1978(E), *Documentation — Numbering of Divisions and Subdivisions in Written Documents* (1978) (see Figure 9.D in Chapter 9).

8. Keep in mind that the report elements discussed within (Abstract, Introduction, Methodologies, Assumptions, and Procedures, etc.) are the recommended minimum elements for scientific and technical reports. Depending on the specific type of report (e.g., equipment safety report, inspection reports, engineering project reports, and progress reports, etc.), other sections may be included if they include critical information. A report dealing with failure analysis, for example, could include a section titled "Prior Failure Histories." Another example is a report that has a section on "Specification Requirements" in situations where "code and specification compliance, nonconformance findings and resolutions" (Thielsch 1996, 5) were of major concern.

Figure 8.H.2 Sample Summary and Conclusions (Page 2 of 2) of a Report From the U.S. Nuclear Regulatory Commission

SUMMARY AND CONCLUSIONS   Section 6

(3) The through-wall cracks in the STS410 carbon steel pipe experiments all grew out of the circumferential crack plane, much in the same manner as the through-wall cracks tested in the U.S. manufactured carbon steel pipe materials evaluated in the Degraded Piping, IPIRG-1, and Short Cracks programs. Such behavior is typically attributed to toughness anisotropy effects. All information to date suggests that the angular crack growth increases the load-carrying capacity and overall ductility (global displacements and pipe rotation), and hence is desirable.

9. Use tables, graphs, and other graphics to support complicated textual material.

10. Traditionally, reports have been written in the passive voice because most technical professionals believe it is more objective than the active voice. On the other hand, modern writing specialists suggest that the active voice should be used because it is more engaging for the reader. The only problem with using the active voice is that some writers tend to focus more on the "me" rather than on the material covered. The result is that readers become distracted and may overlook critical information. The safest rule to follow is to comply with tradition and use the passive voice; most technical professionals will not criticize the writer for using the passive voice in a report, but they will criticize arrogant and self-centered writing that can evolve through use of the active voice.

11. Ensure that the objectives of the project or study are clear.

Figure 8.1 Sample List of References of a Report From the National Aeronautics and Space Administration

REFERENCES

Beckman Instruments, Final Report: Solar Backscatter Ultraviolet/Total Ozone Mapping Spectrometer (SBUV/TOMS) Program, Volume I, Technical Report, 1980, NASA Contract No. NAS5-20970.

Bhartia, P. K., J. R. Herman, R. D. McPeters, and O. Torres, 1994, "Effect of Mount Pinatubo on Total Ozone Measurements From Backscatter Ultraviolet (BUV) Experiments," *J. Geophys. Res.*, 98, 18547-18554.

Bhartia, P. K., S. Taylor, R. D. McPeters, and C. Wellemeyer, 1995, "Application of the Langley Plot Method to the Calibration of the Solar Backscattered Ultraviolet Instrument on the Nimbus-7 Satellite," *J. Geophys. Res.*, 100, 2997-3004.

Cebula, R. P., H. Park, and D. F. Heath, 1988, "Characterization of the Nimbus-7 SBUV Radiometer for the Long-Term Monitoring of Stratospheric Ozone," *J. Atm. Ocean. Tech.*, 5, 215-227.

Fleig, Albert J., Pawan K. Bhartia, Charles G. Wellemeyer, and David S. Silberstein, 1986, "Seven Years of Total Ozone From the TOMS Instrument—A Report on Data Quality," *Geophys. Res. Lett.*, 13, 1355-1358.

Fleig, Albert J., R. D. McPeters, P. K. Bhartia, Barry M. Schlesinger, Richard P. Cebula, K. F. Klenk, Steven L. Taylor, and D. F. Heath, 1990, "Nimbus-7 Solar Backscatter Ultraviolet (SBUV) Ozone Products User's Guide," *NASA Reference Publication, 1234*, National Aeronautics and Space Administration, Washington, DC.

Gleason, J. F., P. K. Bhartia, J. R. Herman, R. McPeters, P. Newman, R. S. Stolarski, L. Flynn, G. Labow, D. Larko, C. Seftor, C. Wellemeyer, W. D. Komhyr, A. J. Miller, and W. Planet, 1993, "Record Low Global Ozone in 1992," *Science*, 260, 523-526.

Heath, D. F., A. J. Krueger, H. R. Roeder, and B. D. Henderson, 1975, "The Solar Backscatter Ultraviolet and Total Ozone Mapping Spectrometer (SBUV/TOMS) for Nimbus G," *Opt. Eng.*, 14, 323-331.

Heath, D. F., A. J. Krueger, and H. Park, 1978, "The Solar Backscatter Ultraviolet (SBUV) and Total Ozone Mapping Spectrometer (TOMS) Experiment," *The Nimbus-7 User's Guide*, edited by C. R. Madrid, pp. 175-211.

Herman, J. R., R. Hudson, R. McPeters, R. Stolarski, Z. Ahmad, X.-Y. Gu, S. Taylor, and C. Wellemeyer, 1991, "A New Self-Calibration Method Applied to TOMS/SBUV Backscattered Ultraviolet Data To Determine Long-Term Global Ozone Change," *J. Geophys. Res.*, 96, 7531-7545.

Herman, J. R. and D. Larko, 1993, "Ozone Depletion at Northern and Southern Latitudes Derived From January 1979 to December 1991 Total Ozone Mapping Spectrometer Data," *J. Geophys. Res.*, 98, 12783-12793.

Hilsenrath, E., R. P. Cebula, M. T. DeLand, K. Laamann, S. Taylor, C. Wellemeyer, and P. K. Bhartia, 1995, "Calibration of the NOAA-11 Solar Backscatter Ultraviolet (SBUV/2) Ozone Data Set From 1989 to 1993 Using In-Flight Calibration Data and SSBUV," *J. Geophys. Res.*, 100, 1351-1366.

Hughes STX Corporation, 1994, "Nimbus-7 TOMS Wavelength Scale Adjustments," *Hughes STX Doc. HSTX-3036-212-MD-94-007.*

Jaross, G., A. Krueger, R. P. Cebula, C. Seftor, U. Hartmann, R. Haring, and D. Burchfield, 1995, "Calibration and Post-Launch Performance of the Meteor-3/TOMS Instrument," *J. Geophys. Res.*, 100, 2985-2995.

Komhyr, W. D., R. D. Grass, and R. K. Leonard, 1989," Dobson Spectrophotometer 83: A Standard for Total Ozone Measurements, 1962-1987," *J. Geophys. Res.*, 94, pp. 9847-9861.

McPeters, R. D., A. J. Krueger, P. K. Bhartia, J. R. Herman, A. Oaks, Z. Ahmad, R. P. Cebula, B. M. Schlesinger, T. Swissler, S. L. Taylor, O. Torres, and C. G. Wellemeyer, 1993, "Nimbus-7 Total Ozone Mapping Spectrometer (TOMS) Data Product's User's Guide," *NASA Ref. Publ. 1323*, National Aeronautics and Space Administration, Washington, DC.

McPeters, R. D., et al., 1996, "Nimbus-7 Total Ozone Mapping Spectrometer (TOMS) Data Product's User's Guide," *NASA Ref. Publ. Planned*, National Aeronautics and Space Administration, Washington, DC.

Paur, R. J., and A. M. Bass, 1985, "The Ultraviolet Cross-Sections of Ozone: II. Results and Temperature Dependence," *Atmospheric Ozone*, Edited by C. S. Zerefos and A. Ghazi, 611-616, D. Reidel, Dordrecht.

Schlesinger, B. M. and R. P. Cebula, 1992, "Solar Variation 1979-1987 Estimated From an Empirical Model for Changes With Time in the Sensitivity of the Solar Backscatter Ultraviolet Experiment", *J. Geophys. Res.*, 97, 10119-10134.

Schoeberl, M. R., P. K. Bhartia, E. Hilsenrath, and O. Torres, 1993, "Tropical Ozone Loss Following the Eruption of Mt. Pinatubo," *Geophys. Res. Lett.*, 20, 29-32.

Stolarski, R. S., R. Bojkov, L. Bishop, C. Zerefos, J. Staehelin, and J. Zawodny, 1992, "Measured Trends in Stratospheric Ozone," *Science* 256, 342-349.

Willson, R. C., H. S. Hudson, C. Frolich, and R. W. Brusa, 1986, "Long-Term Downward Trend in Solar Irradiance," *Science 234*, 1114-1117.

12. Ensure that all data are properly documented.

13. Ensure that your conclusions are supported by the data.

14. Check that the format (placement of headings and subheadings, presentation of tables, charts, and other graphics) is consistent throughout the report.

15. Have a colleague or supervisor review the report. Such an individual should not hesitate to be candid. Remember that your reputation, and perhaps that of your company or organization, may be at stake. It is best to have a colleague tell you that your conclusions or recommendations are

Figure 8.J Sample Errata Sheet

ERRATA SHEET

Please note the following changes to the
report on Thermal Barrier Coating Analysis:

Change 1

Reads

order incrementally from 2.5 to 3.8 demonstrates

Should Read

order incrementally from 2.8 to 3.9 demonstrates

Change 2

Page Number and Line Number

5, 25

Reads

In fact, for the mathematical order to 2.5, the

Should Read

In fact, for the mathematical order equal to 2.8, the

incorrect than to have a client tell you.

16. An errata sheet should be used to clarify information or to correct incorrect information found in a report after printing or even after it has been distributed. Errata sheets should only be used when there are major misunderstandings or errors. They should not be used to correct simple grammatical or spelling mistakes unless such errors change the intended meaning of the text. The errata sheet should specify the page and line number of the mistake, include the incorrect material, and specify the correct words, spelling, or meaning (see Figure 8.J).

Conclusion

Following the recommendations provided in this chapter and paying proper attention to the mechanics of writing will help the technical professional develop technical and sceintific reports that incorporate all required elements and meet national and international standards. Specifically, these are standards developed to ensure that report material is presented to the reader in a format that is logical, consistent, and easy to follow. The results will be development of a report that is easy to read, attractive, and a direct reflection of the writer's skill and professionalism. More important, these recommendations provide for a report where the writer is successful in documenting and communicating the intended message or information to the reader or target audience.

PROCEDURES

A common practice for many professionals is to begin collecting procedures from plants and other facilities they visit. Most organizations, unless they are working for a federal agency, are willing to share their procedures with others. In fact, it is safe to say that most of today's procedures are a combination of original work and a large portion of work (e.g., format, style, and even verbiage) of procedures from other organizations. As most professionals will be faced with the task of writing procedures in their careers, obtaining samples along the way in order to build a procedure library is a good practice to follow.

Introduction

Technical procedures are a "means or method by which action shall be taken consistent with applicable principles; a means of implementing policy" (Cleland and Kerzner 1985, 187) and are some of the most critical documents that technical professionals will be required to write. The development and implementation of technical procedures (hereafter called procedures) help ensure that equipment is properly operated and that manufacturing and service processes are properly conducted. The result is that all operations and services are conducted in a standardized fashion in the interest of safety and quality of standards. Also, utilization of procedures can help organizations maintain cost controls so that profits are maximized.

Although the specific format of procedures will vary from company to company, they all have essentially the same elements and are generally developed and implemented in the same fashion. The technical professional who has written and updated procedures for one company will have the appropriate experience to write and update procedures at another company. This will be true regardless of whether the company is a nuclear power plant or a manufacturing plant that produces components for automobiles or sophisticated telecommunications equipment.

The remaining portions of this chapter provide information on the process of developing, implementing, and updating procedures in engineering, scientific, or technical settings. Also, several examples will be provided from two different types of organizations.

Anatomy of a Procedure

Each company or organization will usually have its own format for procedures. Even with this potential diversity, almost all organizations use at least seven common elements that constitute a procedure; this is supported by two examples of procedures in Figures 9.A.1 through 9.A.6, as well as 9.B.1 and 9.B.2. Both procedures contain the minimum elements required: (1) Purpose, (2) Responsibilities, (3) References, (4) Procedure, and (5) Related Forms and Documents.

The specific elements and definitions for these procedure elements, along with a notation of whether the respective element is "required" or "recommended," are as follows:

1. ***Banner and Approval Plate (required)*** — The banner and approval plate are usually placed on the first page or can also serve as the cover page. The banner includes five subelements: (1) company or organization name; (2) procedure title; (3) procedure number based on a company-wide procedural numbering system; (4) effective or revision date; (5) page numbering that includes the specific number of pages and specific page number (e.g., page 1 of 2, page 2 of 2, etc.); and (6) name and signatures required for appropriate approval of the respective procedure.

   The use of authorizing name and signature on all procedures is important for several reasons. First, it helps ensure that all department heads and other critical management personnel have read and approved the procedure. By placing their names and signatures on procedures, these managers are asserting that to the best of their knowledge the respective procedure is in compliance with all pertinent industry safety standards and practices. Also, use of the authorizing names and signatures ensures that critical personnel are not left out of the procedure review and approval process.

   Although the banner would be placed on every page of the procedure, the names and authorization signatures usually appear only on the first page (see Figure 9.A.1— Page 1 of 6).

2. ***Purpose or Scope (required)*** — The purpose or scope specifies exactly which operations or processes the procedures must be used for. In most cases, this section will be a short paragraph with no more than a few sentences (see Figure 9.A.1— Page 1 of 6).

3. ***Responsibilities (recommended)*** —The responsibilities specify titles of all personnel who will be responsible for partial or total compliance with the procedure. Only titles, not individual names, should be included in this section. Again, as will be discussed in a later section, specifying a personnel title means that the respective procedure must be updated if one or more of the titles specified are changed or eliminated at the respective company (see Figure 9.A.1— Page 1 of 6).

4. ***Reference (required)*** — The references section includes all source document materials specified or cited in the procedure, as well as all

Figure 9.A.1 Sample Procedure Number 1 — Page 1 of 6

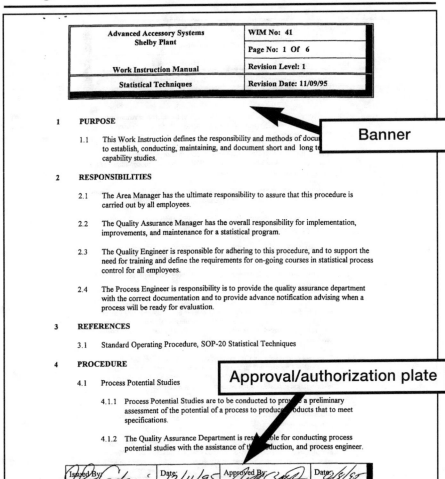

Advanced Accessory Systems Shelby Plant	WIM No: 41
	Page No: 1 Of 6
Work Instruction Manual	Revision Level: 1
Statistical Techniques	Revision Date: 11/09/95

1   PURPOSE

1.1   This Work Instruction defines the responsibility and methods of docu[ment] to establish, conducting, maintaining, and document short and long t[erm] capability studies.

Banner

2   RESPONSIBILITIES

2.1   The Area Manager has the ultimate responsibility to assure that this procedure is carried out by all employees.

2.2   The Quality Assurance Manager has the overall responsibility for implementation, improvements, and maintenance for a statistical program.

2.3   The Quality Engineer is responsible for adhering to this procedure, and to support the need for training and define the requirements for on-going courses in statistical process control for all employees.

2.4   The Process Engineer is responsibility is to provide the quality assurance department with the correct documentation and to provide advance notification advising when a process will be ready for evaluation.

3   REFERENCES

3.1   Standard Operating Procedure, SOP-20 Statistical Techniques

4   PROCEDURE

4.1   Process Potential Studies

Approval/authorization plate

4.1.1   Process Potential Studies are to be conducted to pro[vide] a preliminary assessment of the potential of a process to produce [pr]oducts that to meet specifications.

4.1.2   The Quality Assurance Department is res[ponsi]ble for conducting process potential studies with the assistance of t[he] [pro]duction, and process engineer.

Issued By: _____ Date: 2/11/95   Approved By: _____ Date: 2/8/95

CONTROLLED DOCUMENT

supplementary documents (including other procedures) that should be used together with the respective procedure. In all instances, the latest versions of the references must be used. Specifying the reference publication dates means that the respective procedure must be updated once a new revision or version of the source document is released. Only documents that are directly associated with the respective procedure should be cited in the reference section (see Figure 9.A.1— Page 1 of 6).

5. *Definitions (recommended)* — The definition of all technical terms and titles used in the procedure may be included. Usually, this section follows the

Figure 9.A.2 Sample Procedure Number 1 — Page 2 of 6

Advanced Accessory Systems Shelby Plant	WIM No: 41
	Page No: 2 Of 6
Work Instruction Manual	Revision Level: 1
Statistical Techniques	Revision Date: 11/09/95

4.1.3 Characteristics are to be selected on the basis of the customer requirements, during AQP review, known process variables or other appropriate criteria, e.g., FMEA's, and or D.O.E..

4.1.4 A study is to be conducted by the quality assurance department on the initial production of new or revised processes. With the processing conditions of the sample run to be controlled and stabilized.

4.1.5 Studies are conducted using variable data on a sample quantity of a minimum of 30 pieces of each cavity from a production run of at least 300 shots.

4.1.6 The data is gathered in sub-groups of consecutive units (typically 5 units per sub-group) from the production run.

4.1.7 The data is analyzed using capability analysis report or other appropriate charts, which show no points out-of-control or evidence of negative trends.

4.1.8 Samples are to be checked using measuring devices with approved gage R & R of 10% maximum.

4.1.9 The minimum acceptable outcome for preliminary capability studies, for normally distributed process these requirements are equivalent to a Ppk of 1.67. On an exception basis, preliminary capability levels between 1.33 and 1.67 may be approved in a Inspection Control Plan by the Customer quality activities.

4.1.10 Processes exhibiting a Cp/Cpk value of less than 1.0 require 100% inspection. The Quality Engineer will inform the Quality Manager when a process is calculated to be less than 1.0 Cp/Cpk. 100% inspection is to continue until the re-calculation indicate that the process is within acceptable limits.

4.1.11 Capability studies are to accompany all PPAP submissions. Studies must document Cp and Cpk values taken from an initial 300 piece run using production tooling.

4.1.12 Suppliers are required per the Supplier Requirements Manual, to provide evidence of process capability of selected critical characteristics by the project team, for PPAP submissions and on a quarterly basis for on going capability.

reference section. The definitions can be locally developed or standardized industry terms that would be included in industry codes and standards.

6. *Procedures (required)* — The procedures specify the exact sequences and steps or tasks that must be followed for the respective operations or processes (specific information on writing the procedures will be provided in a later section of this chapter) (see Figures 9.A.1– 9.A.5— Pages 5 and 6).

7. *Attachments (recommended)* — This section includes all forms, tables, and figures referenced in the procedure. Providing the referenced forms, tables, and figures as part of the procedure helps ensure that older versions are not being used in the field (see Figure 9.A.1— Page 6 of 6).

Figure 9.A.3 Sample Procedure Number 1 — Page 3 of 6

Advanced Accessory Systems Shelby Plant	WIM No: 41
	Page No: 3 Of 6
Work Instruction Manual	Revision Level: 1
Statistical Techniques	Revision Date: 11/09/95

4.2   ONGOING STATISTICAL PROCESS CONTROL AND MAINTENANCE

4.2.1   When process capability has been established, evidence of ongoing SPC must be provided. The following methods are acceptable:

    4.2.1.1   Average and Range (X & R) Charts (Variable Data)
    4.2.1.2   Median and Range (X & R) Charts (Variable Data)
    4.2.1.3   C Charts, P Charts, U Charts
    4.2.1.4   Charts for individuals

4.2.2   Key factors / characteristics are to be reviewed from the study for statistical monitoring (charting) as a means of "prevention" for re-occurring non-conformities, centering, the process and reducing variation.

4.2.3   Shifts in process capability will be the determining factor in control limit recalculation frequency.

4.2.4   When an operation is changed to accommodate a new part run, the assigned production supervisor or quality inspector is to ensure the appropriate instruction sheets, (ICP), X-R chart(s), and gages are available to the operator.

4.2.5   The Quality Assurance Department will maintain master charts and will be the source of chart supply and replacement.

4.2.6   Xbar & R control charts is to be completed by the assigned operator.

4.2.7   The operator is to record measurements (typically 5 piece sample subgroups). Each subgroup will be summed (added), then averaged (X), and the range determined (High-Low). Calculations will be recorded and notes pertinent to that subgroup, added to the back side of the chart.

4.2.8   The production supervisor and or quality inspector is to inspect, evaluate and initial each SPC chart used during his/her shift. The supervisor and or inspector is to review the following:

    4.2.8.1   The correct control chart for the part being processed.
    4.2.8.2   The chart is filled out correctly (i.e. enough checks, no biased data and correctly plotted).

Steps in Writing a Procedure

As illustrated in the flow chart in Figure 9.C, a minimum of 21 steps are used when writing a technical procedure and these are subdivided into four stages: (1) Research and Development; (2) Validation; (3) Verification; and (4) Production. Following these steps will help ensure that the procedure is comprehensive and accurate and that all responsible and affected personnel have had input into the development of these critically important documents. These stages and steps are as follows:

1. Research and Development
    a.   Contact colleagues working at other companies similar to your own

Figure 9.A.4 Sample Procedure Number 1 — Page 4 of 6

Advanced Accessory Systems Shelby Plant	WIM No: 41
	Page No: 4 Of 6
Work Instruction Manual	Revision Level: 1
Statistical Techniques	Revision Date: 11/09/95

       4.2.8.3   Reaction to out of control conditions in a timely manner.
       4.2.8.4   Notes recorded for any and all changes in the process.
       4.2.8.5   Trends, Runs, Points out of control.

4.2.9 The operator is to notify the production supervisor and or inspector for direction after recalculation confirmation of all abnormal process variations. See Parg. 4.3 ACTIONS REQUIRED FOR OUT-OF-CONTROL PROCESSES

4.2.10 The production supervisor and or inspector is to inform the Quality Engineer of abnormal process variations.

4.2.11 Adjustments made by the Quality Engineer is to be detailed in the process monitoring log. The Quality Engineer is to notify the production supervisor of the adjustments made, and document the adjustments on the reverse side of the chart.

4.3   ACTIONS REQUIRED FOR OUT-OF-CONTROL PROCESSES

4.3.1 The purpose for actions on out-of-control processes is to establish uniform documentation and actions required in response to charted processes indicating out of control / abnormal variation.

4.3.2 The action required is to be the joint responsibility of the Manufacturing, Engineering, and Quality Assurance Departments.

OUT-OF-CONTROL A process is said to be out of Statistical Control when:

4.3.3 The Control Chart shows the presence of one or more points beyond the control limits for averages (X) or median (X), and the range (R).

4.3.4 Six (6) points on one side of the center line for the averages (X) or median (X).

4.3.5 Six (6) successive points constantly increasing (going up) on the averages (X) or median (X).

4.3.6 Six (6) successive points constantly decreasing (going down) on the averages (X) or Median (X).

to get copies of related procedures. As some procedural steps will be unique for each organization, these copies should only be used for guidance to help get the process going. Many professionals have learned the hard way that adoption and transplantation do not always work. Having samples from other organizations will prove to be a tremendous help in terms of both format and content.

A common practice for many professionals is to begin collecting procedures from plants and other facilities they visit. Most organizations, unless they are working for a federal agency, are willing to share their procedures with others. In fact, it is safe to say that most of today's procedures are a combination of original work and a

Figure 9.A.5 Sample Procedure Number 1 — Page 5 of 6

Advanced Accessory Systems Shelby Plant	WIM No: 41
	Page No: 5 Of 6
Work Instruction Manual	Revision Level: 1
Statistical Techniques	Revision Date: 11/09/95

IN-CONTROL A Process is said to be In Statistical Control when it passes all of the above criteria.

4.3.7   If it is found by the Operator that the process is out-of-control (as indicated in the definition), the following responses will be made:

4.3.7.1   The quality inspector is to verify the operators results. If results of out-of-control is verified. The operator is to resample the process to confirm the out of control situation.

4.3.7.2   If it is shown that the process is indeed out-of-control, the operator is to notify the production supervisor for direction.

4.3.7.3   The operator or production supervisor is to document the actions taken on the back of the X & R Chart. The production supervisor is to initial the action statement to indicate verification. Cpk values are to be included in the documented explanation of actions taken.

4.3.7.4   If it is determined that 100% inspection is required, the operator is to identify and segregate all suspect parts produced since the last in control point by means of a red Do Not Use Tag.

4.3.7.5   The production supervisor is to implement 100% inspection (method and manpower) as required.

4.3.7.6   If a trend indicating a favorable shift in the process (Cpk) continues for two (2) successive shifts (sixteen (16) continuous points), the production supervisor and/or inspector is to contact the Quality Engineer and request a control limit adjustment to reflect the favorable shift. The Quality Engineer is responsible for implementing justified changes.

4.3.7.7   Process driven control limit adjustments is to be highlighted on the existing chart in a contrasting color to define the difference between the old and new control parameters. The production supervisor and/or quality inspector is to document the change on the reverse side of the chart indicating time and reasons for the change.

large portion of work (e.g., format, style, and even verbiage) of procedures from other organizations. As most professionals will be faced with the task of writing procedures in their careers, obtaining samples along the way in order to build a procedure library is a good practice to follow (see Chapter 13 for more recommendations related to the technical professional's library).

b. Collect all source documents related to the respective procedure. Source documents will include all current industry codes and standards, federal regulations, state regulations, equipment operations and technical manuals, vendor instructions, and any memos or other company correspondence related to the respective procedure. It is

Figure 9.A.6 Sample Procedure Number 1 — Page 6 of 6

Advanced Accessory Systems Shelby Plant	WIM No: 41
	Page No: 6 Of 6
Work Instruction Manual	Revision Level: 1
Statistical Techniques	Revision Date: 11/09/95

5 RELATED FORMS AND DOCUMENTS

5.1 First Piece Inspection Control Plan
5.2 Capability Studies, form WIM41-1
5.3 Inspection Control Plan
5.4 X Bar And R Chart
5.5 Red Do Not Use Tags

critical to ensure that the latest edition or versions of the source documents are used.

c. Based on a review of the source documents, determine the person responsible for writing the respective procedure. For short procedures, one writer should be assigned the responsibility. If the procedure is long or requires the technical expertise of others, then a head writer should be appointed. The head writer will be responsible for the coordination of all stages of the process (including production).

d. Based on a review of the source documents and knowledge of the company's organizational structure, make a list of all employees who are familiar with the processes that the procedure will address.

Figure 9.B.1 Sample Procedure Number 2 — Page 1 of 2

Advanced Accessory Systems Shelby Plant Standard Operating Procedure	SOP: 09B
	Page No: 1 Of 2
	Revision Level: 3
PREVENTIVE MAINTENANCE	Revision Date: 09-14-95

1 PURPOSE

1.1 To document the Company's system for performing preventive, predictive, and scheduled maintenance on equipment that affects the quality of output.

2 RESPONSIBILITIES

2.1 The Maintenance Supervisor is responsible for carrying out preventive maintenance on production equipment.

2.2 The Tooling Engineer is responsible for maintaining tooling.

3 REFERENCES

3.1 Advanced Accessory Systems Quality Policy Manual, Element 4.9.

4 PROCEDURE

4.1 All machines and production tools ("devices") are uniquely identified. Tooling is indelibly marked with identification of its owner.

4.2 Devices are logged into preventive maintenance computer program. This program lists:

4.2.1 ID;
4.2.2 Maintenance tasks and instructions (derived from manufacturer and operator manuals, where relevant);
4.2.3 Time intervals (where relevant)

4.3 Once each month, the program produces a Report specifying the inspection tasks to be carried out on all machines.

Issued By:	Date:	Approved By:	Date:

e. Interview all employees who are familiar with the processes that the procedure will address. Because the interviews can sometimes take an hour or more, it is best to schedule the interviews a few days ahead of time to ensure that the employees can provide quality time for the effort. Also, authorization for the interviews must be obtained by the employees' supervisors. Specific questions should be asked regarding safety issues and practices that should be incorporated. Also, acknowledgment and signatures as proof of the interviews should be obtained from all individuals interviewed.

f. Employees and subcontractors of vendors that provided or installed any equipment associated with the procedure should also be

Figure 9.B.2 Sample Procedure Number 2 — Page 2 of 2

Advanced Accessory Systems Shelby Plant	SOP: 09B
	Page No: 2 Of 2
Standard Operating Procedure	Revision Level: 3
PREVENTIVE MAINTENANCE	Revision Date: 09-14-95

4.4 The Maintenance Supervisor inspects each machine in accordance with the Report. Minor problems, adjustments, etc., are carried out at the time and recorded on the Report; so are defects found that affect the safe operation of equipment. Items requiring repair or lengthy adjustments are recorded on the Report as well.

4.5 The Maintenance Manager reviews and approves the report. The information is entered into the program, resulting in the production of a Work List.

4.6 The Maintenance Engineer carries out all the maintenance and repair tasks specified on the Work List. This work is documented on the List and returned to the Maintenance Manager.

4.7 The results are entered into the computer program for record-keeping purposes.

4.8 Repair activity is recorded in the maintenance system as well. Periodically, maintenance and machine performance is evaluated.

4.9 The Maintenance Engineer maintains inventories of standard repair parts and maintenance substances required for appropriate maintenance of production equipment.

4.10 When new production machinery or tooling is added to the facility, its maintenance requirements are added to the computerized maintenance system and preventive maintenance activities commence as outlined above.

5 **CORRECTIVE ACTION**

5.1 Suggestions for change to this document are processed via SOP-05 (Document and Data Control). Handling or problems arising from activities outlined in this procedure is done in accordance with SOP-14 (Corrective and Preventive Action), as appropriate.

interviewed. Depending on the time frame (i.e., how long has it been since the equipment was installed or modified), these interviews may have to be conducted over the telephone. Specific questions should be asked regarding safety issues and practices that should be incorporated into the procedure. Also, as proof of the interviews, acknowledgment and signatures should be obtained from all individuals interviewed. In the event the interview was conducted by telephone, the signature should be obtained by fax.

g. Based on the review of source documents and interviews, the first draft of the procedure will be developed. A common thread throughout the development of the procedure should be safety and

Figure 9.C Procedure Development Flowchart

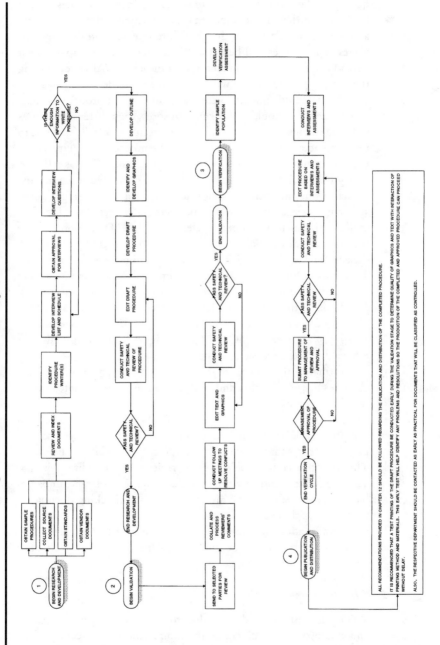

quality. In fact, due to the safety issues, all interview notes should be kept in order to document that all pertinent safety issues were actively considered during the procedure research and development stages. These notes will be helpful if the respective company must defend itself in the event of an accident or liability. After completing the writing procedure, the writer must turn over development notes and all related documents to company officials for proper storage.

h. The procedures should be broken into short, logical steps or tasks as reflected in Figure 9.C. Generally, some specific tasks will have one to two sentences but no more then 10. If 10 sentences are required, then the task should be broken into two or more tasks. Also, as reflected in Figure 9.B.1–9.B.2 the tasks should be presented in an outline format, using numbers for major headings and subheadings. Under such a numbering system, whole numbers (1, 2, 3, etc.) would reflect major topics such as "Scope," "References," and "Definitions," while the decimal side of whole numbers (1.1, 1.1.1, 2.1, etc.) would reflect the actual reference, definition, or tasks and subtasks of the procedure. This number format is used widely in almost all industries for all types of technical documents from technical reports to procedures. Additionally, this numbering format is recommended in ISO 2145-1978(E) *Documentation — Numbering of Divisions and Subdivisions in Written Documents* (ISO 1978) (see Figure 9.D; see also Chapter 8 for a detailed discussion on the ISO).

i. The format of the procedure should, as a minimum, include all the elements discussed earlier and conform to the format adopted by the respective company.

j. All forms, charts, graphs, and other tables associated or referenced in the procedure should be included as attachments. Following this practice ensures that the correct forms, etc., are used in the field.

k. The draft procedure should then be reviewed by an editor who has a strong background in writing and editing. Some large companies have a staff of editors who are responsible for checking documents for proper spelling and grammar.

l. Changes or additions from the editor should then be reviewed by the head writer and incorporated into the next procedure draft; after this, proceed to the validation stage.

2. Validation

Validation consists of sending the draft to all company experts, specialists, and employees for comment on the completeness and accuracy of the respective procedures. The validation stage consists of six steps:

a. Send a copy of the draft procedure, for review, to all company experts, specialists, and employees associated with the respective procedure.

Figure 9.D ISO 2145-1978 (E)

INTERNATIONAL STANDARD **ISO** 2145

INTERNATIONAL ORGANIZATION FOR STANDARDIZATION•МЕЖДУНАРОДНАЯ ОРГАНИЗАЦИЯ ПО СТАНДАРТИЗАЦИИ•ORGANISATION INTERNATIONALE DE NORMALISATION

"The American National Standards Institute (ANSI) is the primary source and official sales agent for ISO standards in the United States. ANSI was granted an exclusive license to distribute and sell ISO standards, technical reports, drafts and other priced publications within the U.S.A. Under this license agreement ISO has granted to ANSI the right to reproduce ISO standards and drafts within the territories of the United States".

Documentation — Numbering of divisions and subdivisions in written documents

Documentation — Numérotation des divisions et subdivisions dans les documents écrits

Second edition — 1978-12-15

UDC 655.535.56 : 002

Ref. No. ISO 2145-1978 (E)

Descriptors : documentation, documents, presentation, numbering.

Price based on 2 pages

ISO 2145-1978 (E)

A cover letter should be provided indicating exactly what is expected from the reviewers. The cover letter should include a statement about ensuring that all related quality and safety issues have been addressed in the procedure. Specify the date by which the comments should be submitted and provide a signoff sheet for comments and signatures verifying review of the procedure.

b. The head writer and team should look at the reviewers' comments and agree on needed changes and incorporate them in the next procedure.

c. Conflicts or discrepancies should be followed up by either the head writer or a member of the procedure writing team if applicable. The key is to go back and meet with any individuals who disagree on any information incorporated in the procedure that either the head writer or writing team members believe is accurate as presented in the first draft. The purpose of the follow-up meeting is to get additional information or clarification on the issue at hand and to reassure the respective individuals that their input is important. This process follows the axiom "Do not ask for advice unless you plan to use it." The follow-up meeting should be documented, along with reasons why the input was or was not used. This documentation may be required to refresh memories in case of a procedure-related accident and legal proceedings.

d. Changes should then be incorporated into the procedure and submitted to the technical editor for editing.

e. Following the editing stage (editing and input of edited changes), another draft should be sent out for comments to all individuals who participated in step b. The same review and input specified in steps b and c should be used for this step.

f. Agreed-upon changes should be incorporated into the document; after this, proceed to the verification stage.

3. *Verification*

The verification stage is used only when the procedures are used by individuals who did not participate in the initial procedure development stage. For example, if a procedure was developed for use by production staff but was developed by the engineering staff, then the verification stage helps ensure that it is written at the proper reading level and that the procedure is clear and functional. There are several key steps associated with the verification stage:

a. Select a sampling of users to read and comment on the procedure. The verification stage is best conducted through an interview process.

b. The verification interview should be conducted on a scheduled basis with the approval of the employee's supervisor and plant or facility manager.

c. The employees should be asked to read the procedure and then explain it to the interviewer. Extreme caution should be taken at this stage so that poor readers are not threatened by the process.

d. During the interview process, a set of standardized, close-ended questions (true or false or multiple-choice answers) should also be used for consistency. The standardized questions should be part of an interview sheet showing the name of the interviewed employee, employee number or security badge number, time and place of interview, as well as the procedure writing team member who conducted the interview. Finally, the employee should be given an opportunity to provide input on how the procedure could be improved for clarification, safety, and quality considerations.

e. Following the interviews the writing team should meet and discuss the results. Use common sense to determine whether changes should be made. The procedure should include any enhancements or language changes that will ensure that all procedure users fully understand the procedure and that there is little likelihood of misinterpretation.

f. All agreed-upon changes and enhancements should be incorporated into the procedure, with a final draft given to the editor for editing.

g. Editing changes should be included in the procedure.

h. The final version of the procedure should then go to the respective managers, administrators, and executives who are responsible for final review and approval of procedures.

i. Any comments or changes from the management review should be incorporated. The procedure should then be sent back to management for final approval. Once the required approvals and signatures have been secured, then the procedure is ready for the production and distribution stage.

4. Production and Distribution

All of the recommendations in Chapter 12 should be followed during this stage. Additionally, consider treating the procedures as "controlled documents." Controlled documents are those that are numbered or serialized with unique distribution or copy numbers so that the management knows exactly who has been provided with a copy. Through a controlled document scheme, holders of the respective controlled document can be immediately notified of any document changes or deletions. Also, the controlled documents may be referred to regularly for a document audit to ensure that the latest version of the documents(s) has been properly inserted in its proper place. The document copy numbering system will help facilitate this process and ensure that all copies have been located.

The controlled document system is used at plants and facilities that operate in critical industries (power plants, solid waste management facilities, water purification plants and systems, etc.) or provide critical services (hospitals, large medical clinics, and biological research laboratories, etc.) where safety to employees and the public are involved. Although it is not required at other types of industries and service organizations, using a controlled document system is recommended because most engineering, scientific, and technical facilities ultimately face many safety issues. Use of a controlled document system can help provide a strong paper and accountability trail that may help show that a company or organization had acted responsibly in case of an accident where an employee or consumer was injured. Granted, the use of a controlled document system will add cost to overhead, but the long-term savings could be millions of dollars if an employee or consumer lawsuit is thwarted.

Additional information on controlled document systems and other records management issues can be obtained from:

Association of Records Managers and
Administrators (ARMA International)
4200 Somerset Drive, Suite 215
Prairie Village, KS 66208
Telephone: 913/341-3808
Fax: 913/341-3742
Web: http//www.arma.org

Revising a Procedure

As illustrated in Figure 9.E, the process of revising an existing procedure is not much different from writing a new procedure. Following the flowchart in Figure 9.E, the only difference is that the first three steps are bypassed in situations where procedures are revised.

Additionally, a critical element in the revision process is to ensure that the revised procedure is sent to staff with explicit instructions that the new document must be inserted in the respective manual and the older version destroyed. As an added step, it is recommended that a transmittal form that employees can sign be sent along with the revised procedures. Signing the transmittal form certifies two facts: (1) that the employee has received, read, and understands the procedure; and (2) that the employee has inserted the revised procedure in the respective manual and has destroyed the older version as instructed.

Having employees — and vendors or contractors doing on-site work — sign the transmittal form with the preceding statements is recommended for two reasons. First, employees will probably take the extra steps to read the procedure when they are required to sign a verification statement. Otherwise, many employees would simply put the revised procedure on their desk, on a pile of other work, with the idea of looking at it later when "things are slower." Unfortunately, in today's work environment where "downsizing" is the norm, things never slow down.

Second, the signed statement provides a strong paper trail that management has taken all prudent actions to ensure that the latest version of corporate or organizational procedures are readily available to all affected employees. Although such proof does not guarantee that a company would not be liable for accidents that could have been prevented if employees had followed proper procedures, it does show that management is responsible and has taken prudent actions to reduce risks or safety hazards to employees and the public. Such proof could, at the very least, show that the company was not negligent and might help persuade the courts and juries to recommend lower fines or damage awards to plaintiffs.

Figure 9.E Updated Procedure Development Flowchart

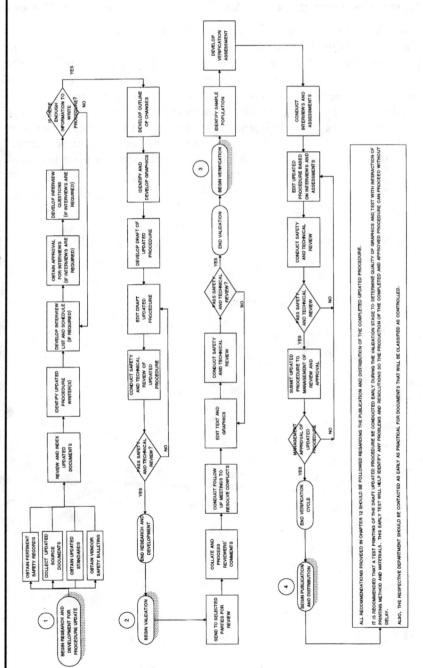

Using the Computer to Assist With the Revision Process

Today's word-processing and desktop-publishing software packages have a "strike-through" capability that can help streamline the revising process. Ideally, the revised procedure should include the new provisions as well as the provisions that are being replaced or eliminated. One way to facilitate this process is to use a "strike-through font" (~~strikethrough font~~) to show the old provisions. To do this, simply select text that will be replaced and then go to the font menu and select the "strike-through" option. The new material can then be inserted using brackets ([ ]). Once the revised document has been approved, all of the strike-through text and brackets must be deleted. The strike-through font and brackets should only be used for minor procedure revisions.

CHAPTER

10

PROPOSALS

"A proposal is a plan of action for fulfilling a need. It is a sales document that is honest, factual, and responsive to the needs of others. It is a written description to be performed that provides enough information for a customer to make a purchase decision"[14] *(Stewart and Stewart 1984, 1).*

R.D. Stewart and A.L. Stewart

Introduction

The process of developing and submitting proposals is common in almost every corner of commerce, from universities to small businesses to large corporations that manufacture sophisticated technology such as fighter jets or personal computers. These proposals are written to obtain funding for research and development, manufacturing of equipment, and even to provide services. Because it is common for the technical professional to write proposals, it is important to master the techniques of writing successful proposals.

In fact, technical professionals who have a proven success record in writing proposals will significantly improve their chances of maintaining employment in these turbulent times of downsizing and corporate merges. Also, having such skills can greatly enhance a technical professional's ability to secure new employment opportunities.

There are two types of proposals that technical professionals may develop during their careers. These types include:

- *Proposals Developed in a Response to Requests for Proposals, Invitations to Bid, or Request for Quotations (Request for Proposal)*
  The most common type of proposal writing occurs when a company is responding to a request for proposals (RFP) or invitation to bid. The RFPs are transmitted by other companies or governmental agencies to companies that have shown an interest in doing business with the company or governmental agency. Sometimes companies may learn of

---

[14] Reprinted by permission of John Wiley & Sons, Inc.

RFPs through information services or announcements published in trade journals or publications. An example would be RFPs offered by the federal government through the *Commerce Business Daily (CBD)* (see Figures 10.A.1 and 10.A.2). The RFP is specific in its technical requirements as well as in the format of the proposal.

- *Proposals Developed for Submission to an Internal Organization or Department*
  In some instances, individual technical professionals or a team may develop proposals for submission to a supervisor, internal organization, or department in response to an internal request for proposals to resolve a problem or improve a manufacturing process or service. Such proposals can also be developed unsolicited, where an employee or team of employees believe that they have a better way of doing something.

No matter which type of proposal you are writing, the purpose is the same: "To develop a product or service in exchange for money which includes external funding or even continuation of funding or employment." In short, the proposal is a sophisticated advertisement of approaches and abilities and is an important part of business in both the private and public sectors.

The remaining portions of this chapter provide information and recommendations to help the technical professional in developing quality proposals. This chapter was specifically designed for development of proposals that will be submitted to an external company or organization, but the same information can be applied to writing proposals that will be submitted internally.

Determining the Odds of Funding

One of the first things that must be accepted by the technical professional is that most proposals are submitted under extremely competitive conditions and only one or a few of the proposals will be funded. Because most of the proposals will not be funded, the technical professional must carefully study the RFP specifications and make three basic determinations:

1. Does the technical professional or company possess the necessary expertise to write a competitive proposal? "Competitive" means that the proposal would score high (in the 90 to 99 range on a 100-point scale). Necessary expertise includes writing skills as well as personnel, or access to such personnel, with the required skills to conduct the project as required or specified in the RFP. The scoring is usually conducted by a peer-review process of corporate or agency executives and technical professionals. For governmental agency grants, the peer-review process is usually conducted by those with technical experience in similar organizations or companies.

Figure 10.A.1 *Commerce Business Daily* Announcement (Page 1 of 2)

INDEFINITE DELIVERY CONTRACT FOR ARCHITECT-ENGINEER (A-E) SERVICES FOR THE ALBUQUERQUE DISTRICT

Category: C Architect and Engineering Services - Construction (PROCUREMENTS)

Date Posted: 1996-06-03

Contact: *US ARMY ENGINEER DISTR, ALBUQ, CORPS OF ENGINEERS, 4101 JEFFERSON PLAZA NE, ALBUQUERQUE NM 87109-3435*

Synopsis:
C -- INDEFINITE DELIVERY CONTRACT FOR ARCHITECT-ENGINEER (A-E) SERVICES FOR THE ALBUQUERQUE DISTRICT SOL DACA47-96-R-0026 DUE 070396 POC Contract Specialist Linda J. Anderson (505) 342-3451 (Site Code DACA47) R-1. CONTRACT INFORMATION: One Indefinite Delivery Contract will be awarded for Architect-Engineer (A-E) services for a period of one year with an option for one additional year. The contract will be primarily for civil works and military projects within the Albuquerque District boundaries, which include Southern Colorado, New Mexico and Southwest Texas, however, the Contractor may be required to work outside this area. The successful firm may be utilized for complete projects or in support of larger in-house efforts. Services during construction to include construction inspection, shop drawing review, and as-built drawing preparation may be required. This Contract will be procured in accordance with Pl 92-582 (Brooks A-E Act) and FAR Part 36. Work will be negotiated and initiated by issuance of delivery orders which will not exceed $150,000. The contract limit is $750,000 per year and the total contract will not exceed $1,500,000. The successful firm is guaranteed no less than $15,000 for the basic year and, if the option year is exercised, the firm shall be guaranteed $7,500 for the option year. When the Government has multiple Indefinite Delivery Contracts for identical services, Delivery Orders will be issued to the various firms based on expertise and workload, provided performance has been satisfactory. The intent is to equally distribute the work. The anticipated Contract award date is December 1996. This procurement is unrestricted. The wages and benefits of service employees (see FAR 22.10) performing under this contract must be at least equal to those determined by the Department of Labor under the Service Contract Act. In accordance with Public Law 95-707, large business firms are reminded that a subcontracting plan will be required which provides for subcontracting work to small and small disadvantaged firms to the maximum extent practicable. If any of the work is to be subcontracted, the Albuquerque District goal is for 60% of the subcontracted dollars to go to small business with 10% and 5% of those dollars going to small disadvantaged business and woman owned business respectively. 2. PROJECT INFORMATION: The work may include engineering studies and reports, preliminary and final designs to include cost estimates, plans and specifications and services during construction. Some designs shall be developed using the

---

2. How many proposals is the funding organization anticipating? More often than not, the organization will have a good idea of how many proposals will be submitted. Sometimes this determination will be based on past experience, or perhaps the company or agency has required companies to submit forms indicating that they plan to submit proposals. Often these organizations will share information on how many proposals they expect.
3. How many proposals will be funded? In manufacturing projects or other private industry initiatives there would be only one contract awarded. In other instances, however, a company may be planning to award more than one contract for a given project.

Based on the answers to the three questions, it must be determined whether a company should go ahead and develop a proposal. Such a determination must be made based on the realization that time is money and that it will cost the company money to develop the proposal. Thus, before taking on the task of developing a proposal,

Figure 10.A.2 *Commerce Business Daily* Announcement (Page 2 of 2)

metric system of measurement. 3. SELECTION CRITERIA: See Note 24 for general selection process. The selection criteria are listed below in descending order of importance (first by major criterion and then by each sub-criterion). Criteria a through e are primary. Criteria f through i are secondary and will be used as "tie-breakers" among technically qualified firms. a. Demonstrated Specialized Experience and Technical Competence in: 1) Air Force Project experience in preparation of plans and specifications and construction cost estimates for administrative offices, building additions/alterations projects, family housing/dormitory projects, aircraft support facilities comprehensive interior design and site development. Firms must also demonstrate engineering design experience with roads, parking areas, utility improvement projects, flood control structures, bridges, and landscape design. 2) The ability to accomplish construction cost estimates utilizing M-CACES Gold or Composer Plus software. 3) The ability to deliver final drawings in AutoCad (Trademark) version 12 usable format and Microstation (Trademark) version 5.0 or later and narratives compatible with WordPerfect (Trademark) format. 4) Implementation of a Design Quality Management Plan. b. Qualified Professional Personnel in the following disciplines: Firms must be able to provide registered professionals assigned in the positions for Project Management, Architecture, Structural Engineering, Mechanical Engineering, Electrical Engineering, Civil Engineering, and Water Resources Hydraulic Engineering. The firm must be able to provide a cost estimator with M-CACES gold or Composer Plus experience. Firms formally trained in using M-CACES is advantageous. The Project Manager(s) must have experience managing design of Air Force and Civil Works projects. The design team must be familiar with relevant Air Force and Civil Works design criteria. c. Past performance on DoD and other contracts in terms of cost control, quality of work, and compliance with performance schedules. d. Capacity to accomplish two concurrent delivery orders. e. Knowledge of the Locality: Firms knowledge of Kirtland, Cannon and Holloman Air Force bases as well as the area encompassed by the Albuquerque District boundaries will be a consideration. f. Location of the firm with respect to Albuquerque, New Mexico, is advantageous provided there are a sufficient number of qualified firms in the area. g. Volume of DoD contract awards in the last 12 months as described in Note 24. h. Extent of participation of SB, SDB, historically black colleges and universities, and minority institutions in the proposed contract team, measured as a percentage of the estimated effort. i. Demonstrated success in prescribing the use of recovered materials and achieving waste reduction and energy efficiency in facility design. In-house capabilities will be weighted heavier than subcontracted work in the evaluation process. Joint Venture Firms will be considered as having capabilities in-house. 4. SUBMISSION REQUIREMENTS: See note 24 for general submission requirements. Interested firms having the capabilities to perform this work must submit one copy of SF 255 (11/92 edition) and one copy of SF 254 (11/92 edition) for the prime firm and all consultants to the above address not later than the close of business on the 30th day after the date of this announcement. If the 30th day is a Saturday, Sunday or Federal Holiday, the deadline is the close of business the next business day. Solicitation packages are not provided. The solicitation number is DACA47-96-R-0026. Only data furnished by the responding firm on the SF 254 and SF 255 will be considered in the selection process. Personnel qualifications of all key subconsultants are to be included in block 7 of the SF 255. Responding firms are requested to clearly identify those tasks to be performed in-house and at what office and those tasks to be subcontracted and at what office. The SF 255 should specifically address the requirements of this announcement. Responding firms are requested to summarize their Quality Management Plan and identify all relevant computer capabilities in block 10 of the SF 255. In addition, responding firms are requested to provide the total amount of DoD contract dollars awarded firm-wide (including contract modifications) for the last 12 months in block 10 of the SF 255. Firms responding as Joint Ventures must state the intent in block 5 of the SF 255. FIRMS SHALL LIMIT BLOCK 10 OF THE SF 255 TO NO MORE THAN 10 PAGES. The POC is Linda J Anderson (505) 342-3451. No other general notifications will be made and no further actions will be required from firms under consideration. This is not a Request For Proposal.(0151)

For assistance in interpreting the *CBD* announcements, please see the *CBD* Reader's Guide.

the company must ensure that it does in fact have the capability to develop a competitive proposal and that competition will not be overwhelming in terms of the number of proposals submitted compared with the number that will be funded. Of course, in situations where only a handful of companies have the expertise and experience to develop the proposal and competently conduct project implementation, the odds are already rather good for those few companies.

Besides the three basic questions, there is one more key question that must be addressed:

Is there a strong possibility that the RFP process is "rigged" or "wired"? In the real world of business, there is always a strong possibility that another company has the inside track on a given project. And, to be perfectly frank, chances are quite high that every company has had, at one time or another, the inside track. The most prevalent situation would be where a company was already doing business for the customer. If this company was doing satisfactory work and had established good working relationships, then chances are that the management of the customer would like to see the other company get additional business. Such situations evolve in both the private and public sectors and do not necessarily mean that anything illegal is going on; it is a part of doing business and must be accepted as a fact of life. Thus, unless a company can prove that another company was awarded a bid because of illegal activity such as payoffs or kickbacks, the only thing that can be done is to accept the fact that another firm has received the award.

This is accepted even in the public sector, where there are strict rules and procedures to ensure that "impartial" awarding of bids is followed. In the public sector there is usually an appeal process for companies and individuals that believe they should have been awarded a project. In only a few instances are the initial decisions reversed; this usually occurs if the complainant can prove fraud or other illegal action. Making such an appeal will usually draw undeserved attention to the complainant, so unless fraud can be shown beyond a doubt, the best advice is to accept the company's decision and move on to the next project.

Preparing to Write the Proposal

Most companies that develop and submit proposals regularly have a procedure or procedures associated with the entire process, from research to writing the proposal to the signatures required before submitting the proposal. Obviously such procedures must be followed, and this chapter will focus on proposal writing and actions that can be taken to help enhance the quality of the proposal. These actions are presented on the assumption that a company or individual has decide to move ahead with the process of developing and submitting a proposal. As reflected in Figure 10.B, the processes associated with developing a proposal should include at least six major steps, which are provided in the remaining portions of this section.

At least six steps must be completed before beginning to write the proposal. These include: (1) Reviewing RFP Specifications; (2) Developing the Plan of Action or Work; (3) Determining Key Project Personnel; (4) Determining Project Costs; (5) Developing Project Schedule; and (6) Developing and Collecting Proposal Drawings and Other Graphics. Each of these steps is discussed here.

1. **Reviewing RFP Specifications.** A great deal of time probably was devoted to the development of RFP specifications and the related statement of work that defines the type of work to be completed. That does not mean, however, that the specifications will be all-encompassing. The first step in developing the proposal is to revisit the specifications to

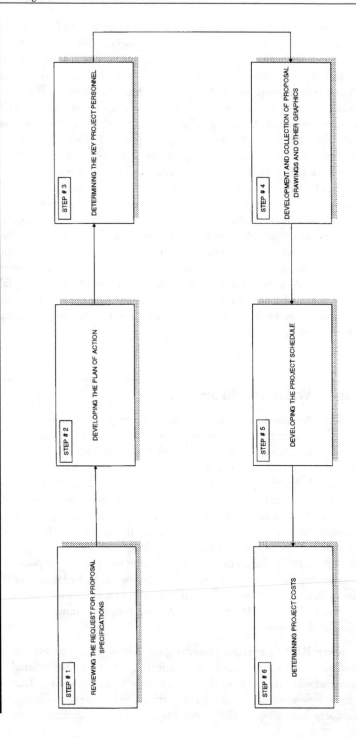

Figure 10.B Proposal Development Flowchart

STEP #1 — REVIEWING THE REQUEST FOR PROPOSAL SPECIFICATIONS

STEP #2 — DEVELOPING THE PLAN OF ACTION

STEP #3 — DETERMINING THE KEY PROJECT PERSONNEL

STEP #4 — DEVELOPMENT AND COLLECTION OF PROPOSAL DRAWINGS AND OTHER GRAPHICS

STEP #5 — DEVELOPING THE PROJECT SCHEDULE

STEP #6 — DETERMINING PROJECT COSTS

Figure 10.C.1 Sample Statement of Work — Page 1 of 5

RFP 10-S-0012-1
ENGINEERING SUPPORT CONTRACT
REQUEST FOR PROPOSAL

SECTION J

ATTACHMENT I

STATEMENT OF WORK

63

determine, to the greatest extent possible, that they are complete and clear. Any unclear wording or passages should be clarified by the designated customer contact. If time permits, such a request should be put in writing and sent via fax, e-mail, or regular mail. Often the customer's response will be sent to the requester and to all other companies that had requested the RFP. A sample statement of work from a RFP from the John F. Kennedy Space Center is provided in Figures 10.C.1 through 10.C.5.

The next step is to develop a plan of action on how to meet the specifications. This process is best conducted by a project team so that the

Figure 10.C.2 Sample Statement of Work — Page 2 of 5

RFP 10-S-0012-1 (ESC)   Section J, Attachment I

STATEMENT OF WORK

1.0 PURPOSE

This Statement of Work (SOW) describes the activities required to be performed by the Engineering Support Contractor (hereinafter referred to as "The Contractor") in order to provide support to the Engineering Development Directorate, John F. Kennedy Space Center, FL. (hereinafter referred to as "DE").

2.0 SCOPE

The Contractor shall be responsible for the performance of the tasks described in this SOW. These tasks will require services which may range in scope from providing technician manpower in support of a variety of sites, facilities, and laboratories to providing engineering and management of complex research, development, and technology projects. These tasks will require the procurement of equipment and systems to be used by the Contractor in support of his continuing design and laboratory activity and the delivery of highly specialized systems to be provided for the use of the government and those other contractors designated by the government. The overall planning, scheduling, prioritizing of Work Orders, and budgeting for these projects is a responsibility of the NASA civil service team with support from the Contractor. Contractor staffing and planning expenditures to meet project requirements are the Contractor's responsibility. Work Orders specifying Contractor support are prepared by Technical Representatives (TR) in accordance with the Work Order procedure set forth in Appendix 3. The Contractor is expected to work with the Contract Technical Manager (CTM) to integrate this support in such a manner that a well-coordinated and consolidated effort results. The Contractor shall make maximum use of available Government equipment, supplies, and services and procure only equipment, supplies, or services which are not readily available and which are required to support accomplishment of assigned tasks. This will include Federal Information Processing (FIP) resources defined in FIRMR (Federal Information Resources Management Regulation) 41 CFR Chapter 201.

3.0 REQUIREMENTS

3.1 GENERAL

The Contractor shall provide the services identified in this SOW.

Government administration and control of the contract will be accomplished through the KSC Contracting Officer (CO). Technical management and direction will be furnished by the CTM appointed by the KSC CO. Technical guidance will be furnished by the TR who will be supported by his designated Technical Contacts (TC).

The Contractor shall provide supervision of its employees and overall direction and integration for all aspects of this contract to ensure that the work activities specified in this SOW are accomplished economically and effectively in compliance with contractual requirements. When directed by the CTM/CO, the Contractor shall establish interfaces to achieve appropriate coordination with NASA/KSC, other KSC contractors, and other NASA Centers. In the event of joint occupancy within a facility or laboratory, the Government shall have the responsibility for prioritizing the use of Government facilities, laboratories, and equipment.

The Contractor shall provide those plans, procedures, and reports identified in the Data Requirements List (DRL) specified in Appendix 1. The Contractor shall establish a resource reporting system, to include subcontractors, for total contract work activity. The reporting

---

plan is balanced and accurate in terms of resources needed to implement it and the technical, scientific, or engineering processes recommended. If the team process is used, there could be considerable disagreement and stress, as all participants will have their own ideas on how the project should be approached. This is normal and will eventually evolve into a dynamic process where all members are contributing in a positive and constructive way. The goal of this process is to reach a consensus on how the project should be approached and who will be the key project personnel in the event that the project is funded. Hopefully, all team participants have had training on the team process and have experience in working in such an

Figure 10.C.3 Sample Statement of Work — Page 3 of 5

RFP 10-S-0012-1 (ESC)   Section J, Attachment I

instructions, System Assurance Analysis (SAA), Operations and Maintenance Documentation (OMD), test procedures, cost estimates, parts lists, software requirements and criteria, software, schedules, environmental data, and other pertinent design data. The Contractor shall conduct periodic design reviews in accordance with DE-P 450, Rev. D, "Design Review Procedure". Design packages shall be checked by the Contractor to assure that technical requirements have been met and editorial errors or omissions have been eliminated prior to approval and release by the Government for implementation.

The Contractor shall thoroughly review designs during the design process and prior to engineering release to assure appropriate safety, quality, maintainability, and reliability requirements are incorporated in the design package. The Contractor shall also be required to develop component, subsystem, and system qualification test requirements, criteria, plans, and procedures.

The Contractor shall provide the necessary development work during the design process, including fabrication, design verification and testing of new or upgraded designs to substantiate the design approach. This effort may also include demonstrations, engineering analysis, and preparation of reports.

The Contractor shall provide and maintain technical expertise in all key technical disciplines for assigned laboratories and other technical areas as may be designated. The Contractor shall maintain expert knowledge and awareness of the current state-of-the-art in these disciplines in industry, academia, and other Government laboratories and programs.

The Contractor shall use off-the-shelf technology to accomplish projects except for advanced development projects where state-of-the-art technology should be used. The Contractor shall develop and perform qualification testing of components not previously approved for use as KSC Ground Support Equipment.

When no appropriate existing technology exists, the Contractor shall provide the necessary scientific and engineering resources to develop the appropriate technology. The Contractor shall devise and perform experiments to validate the chosen technical approach(es) and document all such experiments.

The Contractor shall perform acceptance testing of systems or components developed for the Engineering Development Directorate by the Contractor or by other organizations. The acceptance test procedures and test results will be approved by DE.

The Contractor shall prepare turnover documentation for transfer of operations, maintenance, and sustaining engineering responsibility for completed systems to user organizations.

The Contractor shall perform applications engineering, laboratory tests, and design problems resolution for systems already delivered and turned over to user organizations.

Existing Government-owned computer-aided engineering, computer-aided drafting and other automation tools such as interactive design, graphics and CADnetix systems shall be used as the primary method of operation. However, manual drafting and other methods of documentation shall be used as necessary.

The Contractor shall be responsible for development, documentation, and in-service inspection of all DE responsible pressure systems in accordance with KHB 1710.15B, "KSC Pressure Vessel/System Certification Handbook".

The Contractor shall provide sustaining engineering support for DE facilities, laboratories, systems, equipment, and components as defined by Work Order.

---

environment, which is common in engineering and scientific situations.

In reviewing the RFP, there are several things to look for:

a. Ensuring that specifications are consistent from one item to the next.
b. Determining that the specifications are complete.
c. Determining that the specifications meet all pertinent industry and safety standards.
d. Determining that the specifications include information on how the proposal will be scored. Many proposals will be scored based on a certain number of points allotted to each of the required elements.

Figure 10.C.4 Sample Statement of Work — Page 4 of 5

RFP 10-S-0012-1 (ESC)   Section J, Attachment I

3.3 ENGINEERING SUPPORT

The Contractor shall provide engineering support services which include technical writing and illustrating, engineering drafting, and cost estimating.

The Contractor shall prepare and maintain both formal and informal Engineering Standards, Specifications, and Procedures. This includes a detailed review of NMIs, NHBs, KMIs, and other requirements documents to assure that DE working procedures, standards, and specifications are in compliance and provide recommendations for updating when not in compliance. Services include research, writing, review, editing, typing, and proofreading required to produce a complete document ready for approval and reproduction.

The Contractor shall have a multimedia capability for preparing technical and management illustrations suitable for use in Vu-graph and slide presentations, technical reports, plans, operation and maintenance manuals, training classes, etc. These illustrations shall be sketches, graphs, charts, isometric views, conceptual diagrams, or simplified drawings.

The Contractor shall prepare engineering drawings, block diagrams, schematics, printed circuit layouts, parts lists, layouts, and other associated documentation required for engineering design, studies, criteria, conceptual designs, configuration baselines, and completed engineering activities. All applicable documentation shall be prepared in accordance with GP-435 and KSC-DF-107.

The Contractor shall prepare Drawing Release Authorizations per DE-P 720F, w/changes, "Procedure for Preparation of Document Release Authorization (DRA)", for Government signature prior to release. Engineering Documentation Center services will be provided by the Base Operations Contractor (BOC).

The Contractor shall provide support to the KSC standard parts programs. The effort includes maintenance and generation of component specifications and lists using data sources such as Problem Reporting and Corrective Action (PRACA), NASA Alerts, and Government Industry Data Exchange Program (GIDEP).

The Contractor shall conduct field surveys, gather technical information, and provide operational data to support Government requirements for reviews, special briefings, and investigations.

The Contractor shall provide cost estimating capabilities. These cost estimates will be for new projects, alterations, modifications, contract changes, and studies. All cost estimating tasks shall be accomplished in accordance with KSC-SPEC-G-0002, Rev. B, "Compiling Construction Cost Estimates, Specification for" and KSC SPEC-G-0003, "Ground Support Equipment Cost Estimating, Specification for".

3.4 FACILITIES AND LABORATORIES SUPPORT

The Contractor shall provide technical support to DE facilities and laboratories. These facilities and laboratories are used by various contractor and civil service organizations at KSC for design, development, and research activities. They are also for the Contractor's use in performing research and development engineering, component and qualification testing, and to support tasks described elsewhere in this Statement of Work.

The Contractor shall provide support to facilities and laboratories ranging from operation and maintenance to specific technical services within facilities and laboratories as defined by Work Order. The Contractor shall provide labor, supplies, material, parts, tools, and equipment necessary to sustain the general capability intended for the facilities and laboratories and specific items necessary for particular experimental or test activities. General housekeeping should be practiced regularly to maintain assigned areas in an orderly, professional manner. Periodic calibrations required for test equipment such as volt meters, torque wrenches, pressure gauges, and other laboratory standards are performed by the BOC.

In many cases, the customer will provide information on the scoring process, which is available upon request.

2. **Develop the Plan of Action.** Obviously the plan of action or work (plan of work) will be the very core of the proposal and must be developed in a logical format following project management type task analysis where the entire plan is broken down into small but complementary tasks. Remember that when you write the proposal, the plan of work should include an overall narrative as well as a narrative for each discrete task.

Figure 10.C.5 Sample Statement of Work — Page 5 of 5

RFP 10-S-0012-1 (ESC)   Section J, Attachment 1

The Contractor shall prepare and/or update existing Periodic Maintenance Instructions (PMIs) describing the type and frequency of maintenance to be performed on each of the systems as designated by Work Order.

The Contractor shall provide technical support services for research, design, and development projects and to assure safe and orderly facilities, equipment configuration, and operating practices. The Contractor shall be able to work from preliminary engineering instructions, sketches, and concept descriptions. Technicians shall be capable of using standard machines and equipment normally associated with the disciplines and technologies in each facility and laboratory. The Contractor shall fabricate concept test models, prototypes and limited run production models or modifications thereto as directed by Work Order. Typical functions to be performed are indicated below.

    a.   Plan and coordinate activities.

    b.   Operate, maintain, and repair equipment and tools.

    c.   Fabricate prototypes, models, test articles, and fixtures.

    d.   Perform experiments, tests, and demonstrations.

    e.   Prepare test procedures, record data, analyze test results, and prepare reports.

    f.   Provide technical recommendations and suggestions.

A list of existing facilities and laboratories including examples of the engineering disciplines, technologies, or equipment they contain is provided in Appendix 6. Selected Figures are provided in Appendix 7.

Keep in mind that, if available, the number of points that will be awarded to the plan. Generally, the plan of work will receive the highest or next highest number of points. Obviously, if the plan of work has been allotted a maximum of 50 points out of a total of 100, it should receive the greatest amount of attention during the planning and writing process.

Sometimes, the technical professional may find a better way to approach a project than that presented in the RFP specifications. It is best to present such an alternative approach in an amendment to the proposal or even submit a separate proposal, if allowed. In the public sector, proposals containing alternative approaches are sometimes disqualified. If the technical professional is convinced the alternative approach is the best approach for the customer, then communication with the customer should be established to determine the best and most acceptable approach to presenting an alternative to active legal consideration. Remember the old saying that the customer is always right. If the customer refuses to entertain such an alternative approach, which is certainly the customer's right, then you have two options: (1) develop and submit a proposal following the RFP specifications, or (2) do not develop and submit a proposal. You could also develop and submit the proposal as specified in the RFP and, if awarded the bid, initiate negotiations with the customer to discuss using the alternative approach, which could be less or more expensive. Such negotiations, and even agreements, are acceptable in most private industry situations but are almost never accepted in the public sector.

3. **Determining the Key Project Personnel.** Based on the review of RFP specifications, and input from the project team, the key project personnel should be determined. Since most RFP situations are competitive, it is important to use the best and most qualified personnel for the project. Quality of key project personnel is usually based on professional preparation (education), licenses and certifications held, related experience, and scope of experience. It is best to commit the best and most qualified personnel to the project to help enhance the chances of funding.

In some instances a company may not have individuals with specific skills and experience required for a given project. One way to obtain such expertise is to get a commitment from individuals outside the company that they would be willing to work on the project if it is funded. This practice is common and fully acceptable as long as a notation is included in the proposal that such individuals are not regular employees of the company but are available for the project.

Resumes of the key project personnel selected should be developed in a common format, such as that presented in Figures 10.D.1 and 10.D.2. These resumes will be inserted in the appropriate section of the proposal, as discussed further in this chapter. The information included in Figures 10.D.1 and 10.D.2 includes education and experience in a short, succinct format. Extraneous information such as hobbies, interests, or other non-work-related information is not included in a project resume. In addition, the resume should be developed to match and highlight the education and experience of the key personnel to the education and experience required to competently and efficiently complete the project tasks as specified.

4. **Determining Project Costs.** In proposals dealing with providing goods and services, project costs will generally receive a large portion of the scoring points. Similarly, developing accurate cost estimates for a

Figure 10.D.1 Sample Resume for a Proposal — Page 1 of 2

PHILIP R. PELLETTE

EDUCATION: M.S., Mechanical Engineering, New Mexico State University, 1967
B.S., Math/Physics, Magna Cum Laude, Long Island University, 1962

CORPORATE TITLE: President and CEO

AREAS OF SPECIALIZATION: Project Administration Procedures
Project Planning and Design
Facility Equipment Qualification
Space Vehicle Equipment Testing Program
Aerospace Project Management
Quality Assurance Programs
Training Manual Development
Instrumentation and Control
Records Management Services

YEARS OF RELATED EXPERIENCE: 34

RELATED EXPERIENCE:

Mr. Pellette currently provides, through the firm of Pellette & Associates, technical and project management services to industry and government agencies in several areas.

In his more than 5-year tenure as the Secretary to the senior management safety review board for a commercial nuclear power plant, Mr. Pellette developed the Charter and administrative control methods for that organization, prepared meeting minutes, and established review and to meet Technical Specifications records-keeping procedures requirements.

He was a key figure in the prudence assessment support for a major electric utility. Pellette & Associates personnel, under his direction, prepared and compiled the policy and guidelines which outlined methods and procedures to be followed in accomplishing the tasks for the prudence audit. They prepared the materials used for the initial orientation of the auditors appointed by the Public Service Commission (PSC) and served as the centralized control point for records management including the procedures preparation, compilation, writing/editing, records database development, computer graphics generation, and the review of reports as required in support of the audit.

Additionally, Mr. Pellette and his staff were instrumental in defining, developing, implementing, and operating the computer database used to track prudence assessment activities. These customized data files tracked the assigned responsibilities of the prudence support team and provided daily updates of the status of requested data, responses to key prudence issues, and interview coordination and commitment scheduling.

project can determine whether it shows a profit or loss. In fact some companies have even been forced into bankruptcy for not developing cost estimates accurately. As discussed earlier, there are several excellent publications that focus exclusively on cost estimating, which is outside the scope of this publication. One such publication is *Cost Estimating* (Stewart 1982). Naturally, companies that develop and submit proposals regularly will have one or more technical professionals skilled through experience on cost estimating.

Accurate cost estimating is as much an art as a science because there are

Figure 10.D.2 Sample Resume for a Proposal — Page 2 of 2

Another important service provided by Mr. Pellette and his was the development of training manuals. This service included all aspects of training manual development including research, development, editing, validation, final editing, and production.

Mr. Pellette was Manager of Quality Assurance and developed management systems while at Los Alamos Technical Associates (LATA). Developed and implemented an in-depth quality assurance program for Sandia National Laboratories for the Nevada Nuclear Waste Storage Investigations project for alternative assessment of in-situ storage of high-level radioactive waste. In addition, he was a primary figure in project planning activities and developed the detailed LATA corporate quality assurance program which qualified and certified LATA to perform technical services in compliance with regulatory agencies.

While employed at the Los Alamos National Laboratory (LANL), Mr. Pellette served as the LANL Project Officer for the development of Sprint (W66) and Condor (W73) weapons programs. He coordinated LANL systems development throughout the Laboratory and represented LANL in coordinating activities with the DOE weapons development and manufacturing complex and the military services.

Mr. Pellette was responsible for managing and coordinating the upgrading program of the Los Alamos National Laboratory's safeguards and security alarm system. He planned, scheduled, and managed the tasks involving numerous contractors and suppliers to ensure successful safeguards system installation.

While employed with Grumman Aerospace Corporation, Mr. Pellette was leader of a group performing analog and digital data acquisition during lunar module rocket-development testing. As Staff Assistant to the Base Manager at the NASA White Sands Test Facility, he was assigned as management troubleshooter responsible for ensuring the accomplishment of significant project milestones, including coordination with Sandia Laboratories, the Boeing Tulalip Test Facility, and other testing facilities.

many variables that can affect costs. That is why experience is such an important factor in developing accurate estimates. One such variable is the employees assigned to a specific project. A well-known fact of life in the workplace is that some employees work better and more efficiently than others. The efficiency factor includes both time and the efficient use of resources such as materials and supplies. On larger projects, the cumulative effect of several employees who do quality work but who are not necessarily efficient employees can have a significant negative impact on the profit-loss curve of a project. Such variables cannot be calculated easily by computers or other cost-estimating tools, but an experienced estimator who has worked with the company's employees on past projects can include "employee variable calculations" in the cost-estimating calculations.

Due to the critical nature of cost estimating and its impact on corporate profitability, it is important to have the final project cost estimates reviewed by other employees with such experience. The original cost estimators must be given an opportunity to review the entire proposal before it is submitted to the customer for consideration. There have been instances where the proposal writers have made what seemed to be minor changes to the plan of work after the cost estimators have finished their initial calculations. These minor changes were incorporated into the final proposal and submitted to the customer without having the cost estimators conduct a final check on calculations. However, during the implementation stage of the awarded bid, the minor changes made by the proposal writers turned out to be significant in terms of labor and other costs, which put the project in high cost overruns. Such problems could have been easily avoided it the cost estimators simply had the opportunity to review the final proposal.

5. **Developing the Project Schedule.** Given today's availability of powerful computer workstations and the low cost of project management software such as Microsoft Project and MacProject, there is no plausible reason why project schedules are not developed by computer and incorporated within the proposal. These programs are easy to use and offer the flexibility to develop every type of project schedule from simple Gantt charts (see Figure 10.E) to sophisticated Pert charts.

Thanks to the graphical user interface (GUI) capability of Microsoft Windows, many project management software packages allow the importing or placement of the schedule image directly into the word-processing packages or desktop-publishing packages. This integration is important because, when coupled with all other graphics discussed within this publication, the completed proposal will have a consistent and professional look rather than a mismatch of pages from several programs — which is prevalent even in today's highly technological workplace. The ultimate goal should be to use graphics to develop a sophisticated schedule that can communicate to the customer all of the major tasks and starting and completion dates.

6. **Development and Collection of Proposal Drawings and Other Graphics.** Before even starting the proposal writing stage, it is a good idea to gather all drawings or other graphics that will be incorporated. The preferable format is a computer file so that the drawings and graphics can be integrated into the proposal through the word-processing or desktop-publishing program. Naturally there will be some limitations on what size drawings can be introduced in the proposal as

Figure 10.E Sample Gantt Chart

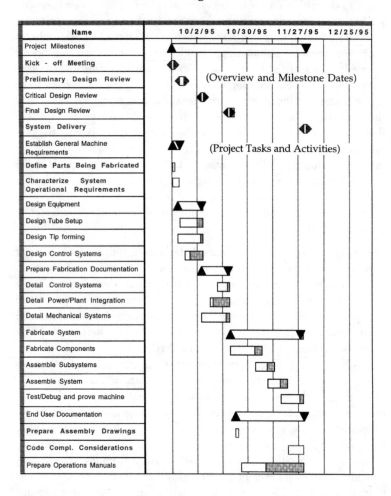

CDS Catheter Forming Machine - Schedule

graphics. Such limitations will depend on how the graphics were created (directly in a computer application program, such as a drafting package or scanned graphics) and how detailed are the drawings. In most situations, you could reduce a "B" size drawing (11 × 17 in.) to fit on a full page as long as the drawing is not too detailed. A power distribution system developed through a computer-aided design package for a "B" size drawing could reduce to a full page of the proposal through the placement and sizing capabilities. A "B" size drawing of a circuit board with several microprocessors, on the other hand, may be too detailed to reduce to a proposal page. Because of such limitations, it is best to have these graphics before developing the proposal. Naturally, if placing graphics and reducing to the normal proposal page size (8 1/2 × 11 in. which presents a small placement area when headers and footers are added) will not work, then the "B" size and larger drawings may be folded and placed in the proposal.

Proposal Format

As specified earlier, the proposal is a sophisticated advertisement of your company. As a result, it is critical that you "put your best foot forward" and develop a high-quality attractive as well as functional proposal. Like any other technical document, the proposal layout must be well planned and presented in a logical format. Typically the RFP specifications will dictate what sections must be included in the proposal. For purposes of illustration the format should include at least five sections:

1. *Identification and Document Organizational Material*
   Some governmental agencies have an identification form that must be included as the very first page of the proposal. The federal government, for example, usually requires Form 424 (see Figure 10.F) to be completed, signed, and inserted as page 1 of most grant and bid proposals. In many instances, these forms are sent out as copies of copies of copies, which have degenerated into unreadable and extremely unattractive documents. Most federal agencies allow the users to develop a computer-generated version of "Standard Form 424." This form, and other customer-provided forms, can be easily produced using a desktop-publishing program. Taking the time to redevelop these forms has three distinct advantages. First, the computer-generated version is more attractive. Second, it is easier to complete with a computer than a typewriter and will be easier to edit. Finally, the computer-generated version can be used with other proposals that may be undertaken in the future. If you plan to develop Standard Form 424 or another cover page form required for a proposal, it is best to check with the customer to obtain authorization.

   Following the form page is the proposal title page. As illustrated in Figure 10.G, the title page traditionally includes the name of the project, name of the customer, name of the company submitting the proposal, and date that the proposal was submitted.

   Next is the table of contents, which should be provided for all proposals over 10 pages. Thanks to today's high-end word processors and desktop-publishing programs, the table of contents can be generated with one key stroke. The attractive feature of the computer-generated table of contents is that it is dynamic and can be easily updated as material is deleted or added to the proposal.

Figure 10.F Standard Form 424

APPLICATION FOR FEDERAL ASSISTANCE	2. DATE SUBMITTED		OMB Approval No. 0348-0043
		Applicant Identifier	

1. TYPE OF SUBMISSION:
Application · Preapplication
☐ Construction · ☐ Construction
☒ Non-Construction · ☐ Non-Construction

3. DATE RECEIVED BY STATE	State Application Identifier
4. DATE RECEIVED BY FEDERAL AGENCY	Federal Identifier

5. APPLICANT INFORMATION

Legal Name: | Organizational Unit: NA

Address (give city, county, state, and zip code): | Name and telephone number of person to be contacted on matters involving this application (give area code)

6. EMPLOYER IDENTIFICATION NUMBER (EIN):

☐☐ - ☐☐☐☐☐☐☐

7. TYPE OF APPLICANT: (enter appropriate letter in box) **[N]**

A. State   H. Independent School Dist.
B. County   I. State Controlled Institution of Higher Learning
C. Municipal   J. Private University
D. Township   K. Indian Tribe
E. Interstate   L. Individual
F. Intermunicipal M. Profit Organization
G. Special District N. Other (Specify) Nonprofit Citizen Organization

8. TYPE OF APPLICATION:
☐ New ☐ Continuation ☐ Revision

If Revision, enter appropriate letter(s) in box(es) ☐ ☐

A. Increase Award B. Decrease Award C. Increase Duration
D. Decrease Duration Other (specify):

9. NAME OF FEDERAL AGENCY:
U.S. Environmental Protection Agency

10. CATALOG OF FEDERAL DOMESTIC ASSISTANCE NUMBER: 6 6 - 8 0 6
TITLE Superfund Technical Assistance Grant

11. DESCRIPTIVE TITLE OF APPLICANT'S PROJECT:

12. AREAS AFFECTED BY PROJECT (Cities, Counties, States, etc.):

13. PROPOSED PROJECT	14. CONGRESSIONAL DISTRICTS OF:	
Start Date	Ending Date	a. Applicant	b. Project

15. ESTIMATED FUNDING:

a. Federal	$	.00
b. Applicant	$	.00
c. State	$	.00
d. Local	$	.00
e. Other	$	.00
f. Program Income	$	.00
g. TOTAL	$	.00

16. IS APPLICATION SUBJECT TO REVIEW BY STATE EXECUTIVE ORDER 12372 PROCESS?

a. YES. THIS PREAPPLICATION/APPLICATION WAS MADE AVAILABLE TO THE STATE EXECUTIVE ORDER 12372 PROCESS FOR REVIEW ON:

DATE _____

b. NO. ☐ PROGRAM IS NOT COVERED BY E.O. 12372
☐ OR PROGRAM HAS NOT BEEN SELECTED BY STATE FOR REVIEW

17. IS THE APPLICANT DELINQUENT ON ANY FEDERAL DEBT?
☐ Yes If "Yes," attach an explanation. ☐ No

18. TO THE BEST OF MY KNOWLEDGE AND BELIEF, ALL DATA IN THIS APPLICATION/PREAPPLICATION ARE TRUE AND CORRECT, THE DOCUMENT HAS BEEN DULY AUTHORIZED BY THE GOVERNING BODY OF THE APPLICANT AND THE APPLICANT WILL COMPLY WITH THE ATTACHED ASSURANCES IF THE ASSISTANCE IS AWARDED.

a. Type Name of Authorized Representative	b. Title	c. Telephone Number
d. Signature of Authorized Representative		e. Date Signed

Previous Edition Usable
Authorized for Local Reproduction

Standard Form 424 (REV. 4-92)
Prescribed by OMB Circular A-102

Then comes a list of illustrations and tables, which should be presented in a format similar to the table of contents.

2. Technical Information or Statement of Work

The technical information of the proposal is the very heart of the document and includes a short introduction to the company and its achievements in related fields.

Figure 10.G Sample Proposal Title Page

A PROPOSAL TO PROVIDE
TECHNICAL WRITING SERVICES

SUBMITTED TO

ENGINEERING SERVICES USA
1042 HARPER AVENUE
HARPER WOODS, MI 48225

SUBMITTED BY

TECHWRITE, INC.
320 PERRY AVENUE
MONROEVILLE, AL 36460

MARCH 22, 1996

Following the introduction is the scope and description of work, which is a detailed description of all tasks and processes associated with the project and how the company will conduct the work. This section must be balanced, provide enough information for the customer to see how the work will be conducted, and demonstrate that the company has a good grasp of the project tasks. Simultaneously, the scope and description of work must not be so detailed that it bores the readers to the point of wasting their time. In short, like other technical writing initiatives, the reader must be engaged and want to read how the work will be conducted.

3. Corporate and Project Management
This section will include a more detailed overview of the company and how it is organized and managed. A specification of previous customers

and related achievements should be provided in here.

Next is a specification of the project management elements, including how the project will be organized and a listing of key project personnel. A project organization chart would be included listing all key project positions. Some organizations provide project-related job descriptions for each key position that will be assigned to the project. This is the section where the resumes of key personnel are inserted.

4. *Project Costs and Payment Schedule*

This section will include the total cost and how the company proposes to be paid for work in progress and the final payment. Naturally, information presentation will be dictated by what the customer requires, as specified in the RFP. Ideally, the company will provide only information that the customer actually requires, as it would not be prudent to include proprietary information such as how the company arrived at a specific cost. In contrast, many governmental agencies require a cost-by-cost breakdown so that proposal evaluators can determine just how the company derived the costs. The premise behind such requirements is the philosophy that companies conducting public-sector work should not generate excessive profits by doing work with governmental agencies. What is excessive is actually, to a certain limit, subjective in nature. Also, many companies have found ways to increase profitability, in a somewhat creative fashion, when doing public-sector work. Contrary to popular belief, such creativity should not be frowned upon because if profits are not acceptable for many companies working in the public sector, then the higher quality companies may simply decide not to do public-sector work. Then, only companies that are not quality organizations and thus cannot obtain private-sector work will gravitate to the public sector. In turn the quality of work received by the public from such companies will be, to say the least, less than desirable.

This section also includes the payment schedule for both work in progress and final payment after an assorted final inspection or sign-off process. Many companies incorporate small discounts if payments are made promptly or early.

5. *Appendices*

The appendices will include all supplementary information such as additional drawings, company financial information, and references. Only required or important information should be included in the appendices. In some competitive situations, information in the appendices will not be considered or even read in determining bid awards. As a result, extreme caution must be taken when choosing what information will and will not be included in the appendices.

How the Proposal Will Be Evaluated by the Customer

Obtaining as much information as possible on how the proposal will be evaluated or scored is extremely critical and can influence how it is written. For obvious reasons, many companies do not "advertise" how they conduct their proposal review process. Based on experience and common industry practices, a few generalizations can be made. First, most companies rely on a team of readers who are technically knowledgeable of the proposal content. Consequently, a large percentage of engineers would be placed on a review team for engineering proposals.

Second, it is common practice to score the proposal using a numerical process where each element in the RFP specifications is given a numeric value. Most often the scores add up to 100 points. How these points are awarded, however, is another matter and can be extremely complicated when lowest costs are not necessarily the deciding factors.

The ideal situation would be to obtain from the customer a specification on how points will be distributed. The practice of providing such information will vary from company to company. The only way that you will know for sure is to ask the customer for scoring information — the most they can say is no. On the other hand, almost all public-sector agencies do provide the scoring information and this is a tremendous asset to the proposal writer. For example, if the scoring system allocates 10 points to the project schedule and 30 points to the costs, then the proposal developers will know that the customer is placing a significant emphasis on costs.

Writing the Proposal

Although much of the initial work associated with developing a proposal generally rests with other professionals who will assist in developing the project tasks, schedules, and cost estimates, the greatest burden will be placed on the technical professional who will be responsible for writing the proposal. If the writer does not do an excellent job of writing the proposal, all of the work conducted by others is academic. In fact, there have been instances when an excellent writer has been able to make up for the poor quality of work provided by the other team members. Unfortunately, it is a rare occurrence when the other team members who have provided quality work can make up for poor-quality work on the part of the writer. In essence, contrary to the old axiom "You should never judge a book by its cover," people do — and the competency of a company will undoubtedly be judged by the cover presented in the form of a proposal. A poorly written proposal will project an image of sloppiness and unprofessionalism, which most customers would rather avoid for purposes of their own survival.

An additional 14 recommendations should be considered when writing proposals. Following these recommendations will, at the very least, help heighten awareness of the writer's goal: to produce a quality document and graphically illustrate to the reader that no other company can do a better job than your company. Equally important is the fact that you must graphically illustrate that your company has the requisite successful experience, personnel, and other resources (equipment, facilities, etc.) to complete the project competently, within budget, and by or before the scheduled completion date. The recommendations are as follows:

1. Be truthful — honesty and integrity are important traits in even the most competitive industries. No company would contract the services of another company that had a reputation of dishonesty — this is especially so with today's corporate liability concerns.

2. Start with a detailed outline listing all major topics and points that must be covered in all sections. Use the outline as a guide through the writing process.

3. Keep the RFP specifications handy during the writing process and refer to them regularly. One of the most common mistakes made in writing proposals is not addressing the RFP specifications, which is the same as not meeting them.

4. The proposal should be written for a technical audience unless it is explicitly known that the proposal will be evaluated by a nontechnical or mixed audience.

5. Use an appropriate level of white space and use headings and subheadings that are descriptive.

6. Avoid the use of humor — proposals are scored on the merit of technical competence, not jokes or interesting antidotes.

7. Be consistent. What you say on one page or section should agree with what you say on another one. Thus, if you say that your company will guarantee the work for five years in one section, then all references to the guarantee in other pages should also specify five years as opposed to three or four years. The guarantee then becomes a thread that runs throughout the proposal. It is extremely important that all such threads remain constant and connected throughout the proposal. It is helpful to write down all proposal threads and double-check them for correctness as you and others proof the proposal.

8. If you experience writer's block on a certain topic or do not have enough information to write on a topic, put a notation in the text in brackets to remind yourself that you need to add some information.

9. Make certain that all project objectives are in measurable terms are clearly presented and meet the RFP specifications.

10 Include only relevant information — readers do not have the interest or the time to go on an excursion.

11. Be concise but thorough.

12. Properly label all illustrations, tables, and other graphics. Make certain that all such graphics are referenced or discussed in the text.

13. Keep the readers in mind — do not make them search for important facts and information. Peers who read and evaluate the proposal appreciate

the fact that the writer has kept them in mind by developing an easy-to-read proposal, as well as one that includes table of contents, table of illustrations and charts, and even indexes.

14. Be realistic — typically, proposals promise too much that can never be delivered. Readers are not fools and will know when they are being given a line. When readers sense that the writer is making promises that are not realistic, they may ultimately turn on the writer and the proposal will not receive a high rating.

Proofing and Editing the Proposal Prior to Production and Submission

Final proofing and editing is a major concern with every type of technical document. For proposals, however, the importance of taking special care in the proofing and editing stage cannot be overemphasized. Unlike technical reports that can be recalled if career-threatening mistakes are found, the submission of a proposal is usually final. Although it may be professionally embarrassing to send out a corrected technical report to cover a mistake in the original version, it could cost a company millions of dollars if a mistake is made in a proposal that meant not getting the contract award.

Due to the importance of proofing and editing a proposal and the business implications, there are six recommendations that should be considered to help ensure that the final version, the one submitted to the customer, is as accurate and competently written as possible. These are in addition to the editing recommendations in Chapter 5. These recommendations are as follows:

1. Identify a senior-level editor who is known, firsthand, to be a competent editor. As discussed in other chapters in this publication, every technical writer must work with an editor. The role of the editor is to ensure that the proposal is grammatically correct and that it is written in a manner that meets all pertinent standards of the respective style manual.

2. Select at least three additional professionals who are intimately familiar with the proposal material. These professionals should be asked to read the proposal and to answer three basic questions:

   a. Are the technical content and approach specified in the proposal correct?
   b. Is the proposal clear and understandable?
   c. Is the proposal realistic?

   The answers should be in a written format and provide recommendations for all "no" responses.

3. All numbers and figures should be double-checked by the individuals who did the original calculations. Have these individuals sign off on the accuracy of the numbers and figures. The practice of asking colleagues

to sign off on projects does not foster trust or warm relationships, but does ensure that they will take a little extra time to ensure accuracy — after all, you are asking them to put their names on the line.

4. Do a hand calculation of all rows and columns of numbers — even if they were calculated by computer.

5. Review the RFP specifications one more time and use a check-off process to ensure that every required specification has been adequately addressed in the proposal.

6. After all these steps and other editing steps (see Chapter 5) have been completed, including making all required changes in the computer, print a final copy and have both the editor and writer read it one more time.

TRAINING MANUALS

*"Effectively designed and written manuals . . . help people do their work
correctly, efficiently, uniformly, and comfortably" (Casady 1992, 17).*
Mona Casady
"The Write Stuff for Training Manuals"
Training & Development

Introduction

There have been two major changes in the workplace during the past 15 years that
have resulted in the critical need to provide comprehensive in-house training programs
for employees: tremendous technological advancements that affected almost every job
function from facilities maintenance and security to production of correspondence
and reports to highly advanced manufacturing processes, and the expansion of
individual work responsibilities, as discussed in earlier chapters. The expansion of
individual work responsibilities means that today's employees must, if they are to
remain employed, work smarter and take full advantage of available technologies to
automate the work process for quality and efficiency purposes.

These two changes alone have increased the need for employee training. Today, business
and industry spend well over $30 billion dollars annually on training employees. A large
portion of these training dollars are devoted to the development and production of
training manuals, which, even in today's video-oriented society, are still a main
component of any training program–whether it is self-paced, computer assisted, or more
structured such as in the traditional lecture-and-discussion format. Unfortunately, a large
portion of these training dollars are wasted because the training manuals developed are
ineffective and **boring** with little incentive to keep the learner actively involved in the
learning process. The result is that these training manuals are rarely used as designed.
Failure to use these manuals and to learn the proper procedures for a given operation
can cost a business millions of dollars due to improper operation and maintenance of
equipment which leads to expensive equipment repairs and causes equipment downtime,
delaying production and delivery of company products and services.

With the preceding in mind, the remaining portions of this chapter focus on processes and procedures that can be followed to ensure that effective, exciting manuals are developed to enhance employee training initiatives.

Hiring External Consultants

One option companies can take to avoid having ineffective and boring training manuals is to contract the services of a technical writing company with *demonstrated successful experience* in developing such documents. Although there are thousands of technical writing companies that claim to be experienced and successful in writing effective training manuals, few can document such success, so caution should be taken when committing financial resources to hiring a consulting team. Such caution can be summarized into four recommendations, as follows:

1. Contact other companies that produce similar goods or services and ask for recommendations on technical writing firms that have developed their training manuals. If the company is cooperative, ask for a copy of one of the training manuals developed by the recommended technical writing firm.
2. Contact the recommended technical writing firms and ask for information on related services and at least five references from companies for which the firm has developed training manuals. There are many technical writing companies that perform a variety of technical writing activities of which writing training manuals is only one such service. Although most technical writing firms may have written one or two manuals during the past year or two, only firms that specialize in training manuals have actually written a large number of manuals. The more experience that firm has in writing training manuals the better, more effective manuals it will produce. If the respective firm has not met this quality goal, then it will not be in business very long. It is extremely important that all references be checked by calling the listed company contact and asking several questions:

   a. What type and how many different manuals were produced?
   b. Were the manuals delivered on time?
   c. Were the manuals provided in a format as originally agreed?
   d. Are employees or customers using the manuals?
   e. What type of feedback have employees or customers provided on the quality and usefulness of the manuals?
   f. Would the company hire the technical writing firm to write other training manuals?

3. Obtain a listing of all technical-writing positions (filled) that the company currently has. Be sure to determine whether these are company employees or contractors who are subcontracted as needed. The preferable configuration is that the company primarily uses its own employees and only hires

subcontractors as needed for special situations. Using employees rather than subcontractors has two advantages. First, the employees have experience in working as a team. As no one employee can have all of the expertise required to write the training manuals on a large-scale basis, a team approach must be used where there are content specialists, writers, illustrators, and editors, all of whom contribute to the development of a quality training manual. Second, the company's ability to hire and support the technical-writing team clearly shows that it has a constant work load. Generally, any company that is able to have a constant work flow is, in all probability, providing quality goods and services to its customers. Finally, the company's mix of employees should include technical writers, training specialists, content specialists, illustrators, and editors that can provide a turnkey operation for the total support of the training manual development and production.

No doubt the smaller "mom and pop" technical writing companies will disagree with this analysis. One argument is that using the smaller technical writing firms for development of training manuals will save a company money. However, this may be a short-term savings because in the long run they cannot provide the breadth of service, including the all-important content area. Using the services of a larger and more experienced company, however, may cost a little more but will help the company save thousands of dollars in training time and help provide better services and products — the long-term savings. This claim is made on the assumption that the respective comprehensive technical writing firm has experience and received high marks from the reference calls. Also, the actual savings, when observed in macro, between using a smaller and less experienced technical writing firm rather than a larger and more experienced firm is quite small. The fee for the development (not production) of the training manual by a large and experienced firm could be, for example, 25 percent higher than the associated fees charged by a smaller firm. The difference between the two fees, for example, would be $12,500. For a company that produces goods and services and generates total sales of, for example, $50 million, the $12,500 difference amounts to only a small percentage of total sales. No doubt, going with a larger company that is more experienced, with an experienced technical-writing team and a proven record of success, will most likely be able to save the company well over the $12,500 in employee time and equipment failure and repairs due to training that resulted, in part, from better training manuals.

There are smaller and newer firms that could develop effective training manuals. The risk factor is significantly smaller in going with a firm that has an experienced team whose success is documented by at least five references. This lower risk factor is critical in terms of saving a company time and money.

4. The initial contract should be for the development of only one manual or module. If the company needs a total of 20 modules, purchasing only one

will cost more than if it were incorporated in a package price of 20 manuals. Having an initial contract for only one manual will provide a chance for the technical writing firm to prove itself and give the company an opportunity to obtain a working sample of the firm's product. If the company is pleased with the first module, then a contract can be issued for subsequent modules. An ideal situation for the purchase of 20 modules, for example, would be to purchase five modules at a time through contracted services. This configuration will put additional pressure on the technical writing firm to provide the best possible manuals so that they will be given the additional contracts. Again, such an arrangement may cost a little more money but will provide the company with a better product, and thus the additional costs will be offset during the implementation of the manuals.

Developing Training Manuals In-House

Another option is to develop training manuals in-house with existing staff. This ability to develop the training manuals in-house is based on several factors. First, qualified staff must be available, including those who are capable technical writers, knowledgeable in technical content, editors, and illustrators. Be sure these staff members are released from their day-to-day responsibilities to devote the proper amount of time to developing the training manuals. Under no circumstances should the responsibility of writing a training manual simply be added to employees' existing responsibilities. Having the required amount of uninterrupted time to properly develop the training manuals will be the difference between success and failure.

The remaining portions of this section focuses on the five basic processes associated with the in-house development of training manuals: (1) Time Required for Initial Research; (2) Research; (3) Development of the Training Manual Outline; (4) Layout of the Training Manual; and (5) Writing of Training Manual Content.

1. **Time Required for Training Manual Research and Development**. Although the time required to conduct research on and develop a training manual will vary from project to project and from employee to employee, there are some industry standards that can be applied to obtain an estimate of how much staff time will be required for this important project. One of the best such standards has been developed by Stewart (1982). As illustrated in Table 11.A, an average of 11.5 person-hours per page is required for developing a training manual. This estimate considers all functions (research, writing, editing, illustration, typing, etc.) required for developing the training manual.

Table 11.A Estimated Person Hours Required for Development of New Training Manual Documentation

Function	Person-Hours Per Page
Research, liaison, technical writing, editing, and supervision	5.7
Typing [word processing] and proofing	0.6
Illustrations	4.3
Engineering	0.7
Coordination	0.2
Total Hours	**11.5**

(A range of 8 to 12 person-hours per page can be used)

Source: Stewart (1982, 111); Reprinted by permission of John Wiley & Sons Inc.

Table 11.B Estimated Person Hours Required for Revising Existing Training Manual Pages

Function	Person-Hours Per Page
Research, liaison, technical writing, editing, and supervision	4.00
Typing [word processing] and proofing	0.60
Illustrations	0.75
Engineering	0.60
Coordination	0.20
Total Hours	**6.15**

(A range of 4 to 8 person-hours per page can be used)

Source: Stewart, (1982, 111); Reprinted by permission of John Wiley & Sons Inc.

Similarly, as illustrated in Table 11.B, revising or updating existing training manuals will require an estimated 6.15 person-hours per page.

The estimates provided in Tables 11.A and 11.B are just that — estimates — but they will be extremely helpful in estimating the total amount of person-hours that will be required to develop or even update training manuals. Through the development of a simple outline, which will be discussed later in this chapter, an estimate can determine approximately how many pages the training manual will be, including the table of contents, content material, illustrations, and index. The total estimated number of person-hours that will be required for the development of the training manual will be derived by multiplying the estimated number of pages of the manual by 11.5 person-hours per page.

2. **Research Stage.** The research stage includes all activities necessary to collect and review all source documents and training manual samples provided by other companies, and conduct employee interviews and on-site reviews. This is a critical stage, which, if conducted properly, will make the entire writing of the training manual a smooth and uneventful process. On the other hand, not taking the necessary time to properly research the training manual topic will mean that the writing process will be a painful and excruciating one. More important, however, not taking the necessary time and steps may mean that critical training information is not included in the manual. Leaving out critical training information can put users in potentially dangerous situations and could cost a company millions of dollars in downtime due to improper equipment maintenance and operation.

With the preceding in mind, the first step of the research stage should be a meeting with trainers and supervisors in the departments that will ultimately be using the training manuals. This meeting should be used to learn as much about the equipment and process as possible and to request copies of all source documentation — vendor technical manuals and company equipment specifications. Also, copies of training manuals previously used by the departments should be obtained. Additionally, all safety-related issues of the equipment operation should be discussed, including obtaining all related industry codes and standards as well as any related standards or regulations issued by the U.S. Occupational Standards and Health Administration (OSHA). Finally, copies of all related company safety violations and records should be obtained.

The next step is to contact the equipment vendor and ensure that the proper manuals are being used. In addition, request a listing of other customers that have purchased similar equipment from the vendor. Finally, send a certified letter to the equipment vendor and request information on all safety issues associated with the respective equipment, including a specification on any accidents or safety concerns that have appeared since publication of the equipment's technical manual(s). The vendor's answer to this letter should be maintained in the project files.

Next, conduct a comprehensive search of literature of related professional journals using key-word searches on the training topic or process. The articles obtained will be an excellent source of overview material, as well as provide some information on the latest industry trends.

3. **Development of the Training Manual Outline.** First, it is important to remember that developing training manuals is significantly different from other types of technical writing. One major difference is

that often the target audiences are not engineers, scientists, or technicians, so training manuals can be written for primarily nontechnical audiences. Next, training manuals are based on specific *"behavioral objectives,"* which specify what the learner is expected to be able to do after completion of the associated training program. The formal definition of a behavioral objective is as follows (Good 1973, 393):

> The aims or objectives of education stated as actual performance criteria or as observable descriptions of measurable behavior.

An example is objectives for a training manual on chemical preparation associated with recycling processes. The behavioral objectives for the training manual would be as follows:

After completing this manual, you should be able to do the following:

- Identify the chemicals used in the recycling process.
- Explain the function of each chemical and the effect it has on the furnish.
- Identify the types of furnish used and explain how each impacts the final product produced on the paper machine (Alabama River Newsprint 1993a,1).

In short, the behavioral objectives specify what the learner will or should be able to do after successfully completing the respective training manual. Once the behavioral objectives have been developed, the next step is to develop an outline for the manual. There are several formats, but the most comprehensive consists of the *Pellette Training Manual Outline*—named after Phillip Pellette, founder of Pellette and Associates, which produced training manuals and other types of technical documents for engineering and scientific organizations. Before starting his own technical writing firm, Phillip Pellette was a mechanical engineer for Los Alamos National Laboratory and Grumman Aerospace Corporation. The Pellette Training Manual Outline consists of a traditional outline structure of sections and subsections of critical elements that form a comprehensive training manual:

- *Table of Contents* — a detailed specification of the major sections and subsections included in the training manual.
- *Introduction* — an introductory section that provides the learner with an overview of the training manual, its objectives, specific resources or references used in developing the module, and a specification of safety-related issues, processes, or procedures.
- *Content Information* — an outline of all major subsections associated with the content information. The outline must flow in a logical direction starting with the most basic concepts and moving toward

the more complicated concepts. The outline for this section should be in a traditional numerical outline format, as follows:

1. First Section Title
   1.1 First Subsection Title
       1.1.1 First Subsection Topic
       1.1.2 Second Subsection Topic
       1.1.3 Third Subsection Topic
   1.2 Second Subsection Title
       1.2.1 First Subsection Topic
       1.2.2 Second Subsection Topic
       1.2.3 Third Subsection Topic
2. Second Section Title
   2.1 First Subsection Title
       2.1.1 First Subsection Topic
       2.1.2 Second Subsection Topic
       2.1.3 Third Subsection Topic
   2.2 Second Subsection Title
       2.2.1 First Subsection Topic
       2.2.2 Second Subsection Topic
       2.2.3 Third Subsection Topic

- *Summary* — a summary of all major points provided in the sections and subsections relating to safety and content information.
- *Progress Review* — a fill-in-the-blank test for the learners to complete at their own pace. The test must be based on safety and content information. An answer key should be provided in the pages following the test.
- *Definitions* — a specification of all industry-related definitions of key words used in the training module.
- *Illustrations* — an inclusion of all illustrations or drawings associated with the respective equipment. Only illustrations that will help learners better understand the equipment should be included.
- *Index* — a specification of all key words and concepts along with page numbers used in the module. Most modern word-processing programs include an index generator.

4. Layout of the Training Manual. Developing a training manual that will be functional, effective, and exciting is the challenge that most technical writers must face. As discussed, the design and layout of a technical document is as important as the content of the training manual. To be sure, if you do not get the reader's attention in the first place, you will never have any opportunity to hold it. Also, as discussed earlier, the layout and design of a technical manual will be a critical ingredient in its readability. Although readability sampling is important, such sampling does not account for the impact that graphics and illustrations have on

**Figure 11.A.1 Title Page from the *Headbox Training Manual*
Developed by Alabama River Newsprint at Perdue Hill, AL**

A L A B A M A   R I V E R   N E W S P R I N T

T R A I N I N G   M O D U L E

Headbox
PMA-05

NOV 0 3 1993

March 25,1991

---

the reader's ability to comprehend the material. A functional training
manual should have good quality content information in text form at
the appropriate reading level and graphics and illustrations to help support
the content information. Also, experience has shown that an effective
training manual must have a good balance of white space to eliminate any

Figure 11.A.2 Sample Introduction Section Page (Page 1) From the *Headbox Training Manual* Developed by Alabama River Newsprint at Perdue Hill, AL

ALABAMA RIVER NEWSPRINT TRAINING MODULE

Headbox

INTRODUCTION

INSTRUCTIONS FOR USING THIS MODULE

- Read and study the material presented. A complete reading from start to finish is recommended and will provide you with the necessary knowledge to operate the headbox.
- Refer to the margin illustrations to aid you in visualizing headbox component locations.
- Study the Illustrations section of this module, referring to the text for explanations.
- New terms are *italicized* in the text, with definitions in the right-hand margin.
- A complete list of terms and definitions is located in the back of this module.
- Reinforce your learning by jotting down notes or questions you may have on the material on the reverse side of each page.
- Demonstrate to your supervisor/instructor your knowledge of headbox operating and adjustment procedures.

OBJECTIVES OF THE MODULE

After completing this module, you should be capable of performing the following:

- Identifying the major component sections of the headbox and their functions.
- Describing the flow process through the headbox.
- Describing the function of a sight glass and its use.
- Describing methods used to control the pressure in the headbox.
- Describing methods used to adjust the direction of flow from the slice into the nip of the wire of the paper machine.
- Explaining the procedure for changing the top slice lip.
- Explaining the procedure for changing micro-adjuster rods.
- Describing the function of the slice heating system.
- Describing the use of the edge feed pipes from the inlet header to the turbulence generator and the slice section of the hea

PMA-05/D-4
Rev. 03-25-91

Computer file name and revision date label

appearance of document clutter. Also, a balance of color is needed to help draw the reader's attention to important concepts, elements, or directives.

Figures 11.A.1 through 11.A.6 show sample pages from the *Alabama River Newsprint Training Manual: Headbox (PMA-05)* (Alabama 1993b). The Alabama River Newsprint Company is located in Perdue Hill, Alabama, and began production in 1992. The six pages from the training module are an excellent example of an attractive and functional document layout that has been able to take full advantage of text, illustrations, and

Figure 11.A.3 Sample Introduction Section Page (Page 2) From the *Headbox Training Manual* Developed by Alabama River Newsprint at Perdue Hill, AL

ALABAMA RIVER NEWSPRINT TRAINING MODULE

Headbox

- Describing the headbox startup and shutdown procedures.
- Describing the method used to clean the headbox.
- Identifying interlocks in the headbox system.
- Identifying lubrication points on the headbox.

RESOURCES
- **Forming Section, P&ID No. F-72-11-001**

SAFETY
Specific safety considerations applicable to operations and maintenance in the headbox area must be adhered to at all times. Operators must be sure to lock and tag out the secondary pump and the primary screen, and make certain the paper machine has been shut down before performing any maintenance on the headbox. The headbox is under pressure and caution should be observed when in this area at all times.

PMA-05/D-4
Rev. 03-25-91

color. As reflected in Figure 11.A.4, the left two-thirds of the document is reserved for the content information, while the remaining one-third is reserved for graphics and industry definitions of key-words introduced in the content information. Also, the page layout incorporates a one-inch header with a line separating the company name and training manual title. As the name implies, the left margin line extends from the left margin to a point of one inch from the right edge of the page.

The next point is the module title (see Figure 11.A.2), which is in initial

Figure 11.A.4 Sample Content Section Page (Page 3) From the *Headbox Training Manual* Developed by Alabama River Newsprint at Perdue Hill, AL

ALABAMA RIVER NEWSPRINT TRAINING MODULE

Headbox

THE PROCESS

The SYM-FLO HS Headbox used at ARN was developed specifically to produce grades of printing paper manufactured on a gap type *former*. The headbox is one of the most important single components affecting final newsprint quality. Headbox performance affects *basis weight profile*, newsprint *profile*, and other general characteristics of the final product. The primary function of the headbox is to transform a circular input flow into a stable rectangular jet stream that flows at a continuous rate over the entire width of the machine. It must also keep the stock properly mixed to prevent air bubbles or dead areas from disturbing the smooth continuous flow onto the wire of the paper machine. The Paper Machine Operator controls headbox adjustments which will determine the final characteristics of the newsprint produced.

former: *section of the paper machine which removes water from the stock to form the fibers into a web of paper.*
basis weight profile: *changes in the basis weight across the width of the sheet.*
basis weight: *the weight, in pounds, of a ream of paper.*
profile: *changes in the caliper (thickness) across the width of the sheet.*

COMPONENTS

Refer to Illustration 1, Headbox.

We will now briefly discuss the individual components of the headbox and their functions. Operating and adjustment procedures will be discussed in the next section.

Inlet Header and Recirculation

The headbox receives accepted stock from the primary screen into a tapered inlet header. The stock must then change direction and flow through a tube bank. A balancing tube with a sight glass is fitted from the back side to the front side of the tapered header for determining the pressure across the header. At no pressure difference, the flow into the headbox from the inlet header is the same from front to back. At the end of the header is a recirculation line, with control valve, which can be opened and closed to adjust the pressure difference across the header. Normally this valve is open to provide approximately 10% recirculation.

During operation the sight glass is filled with stock. If the pressure is higher at one end of the header, the stock will flow through the sight glass toward the lower pressure side. The recirculation line is opened to reduce pressure on the recirculation side of the header, or closed to increase the recirculation side pressure. When there is no movement through the sight glass, the pressure in the header is stabilized.

WHAT IS THE
REAL PURPOSE
OF A HEADBOX...???

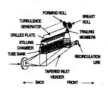

PMA-05/D-4
Rev. 03-25-91

capital letters in boldface type immediately below the left margin line. Next is the section title in all caps, followed by a section title line. Following the section title is the text, which in this case is in 11-point New Times font. Besides section headings, the document also includes subheading titles. Subsection headings and related text are broken down into logical components, which assist in the reader's ability to comprehend the text (see Figure 11.A.3).

Definitions and illustrations are on the right one-third of the pages to

Figure 11.A.5 Sample Content Section Page (Page 14) From the *Headbox Training Manual* Developed by Alabama River Newsprint at Perdue Hill, AL

ALABAMA RIVER NEWSPRINT TRAINING MODULE

Headbox

MAINTENANCE

LUBRICATION
Headbox lubrication will be performed by maintenance personnel.

Headbox components should be lubricated as follows:

1. Worm gear jack for vertical adjustment of the top slice
   - Lifting screws
     - Use 0.05 kg of lubricant type A every 6 months.
   - Worm gears, KV 85-24KE
     - See Katsa maintenance instructions.
   - Universal shafts
     - Use 0.05 kg of lubricant type A every 6 months.
   - Gear motor, Bauer DP2A122CFG1-171/309-MG
     - See Bauer maintenance instructions.
2. Worm gears for horizontal adjustment of top slice
   - Transfer screws
     - Use 0.05 kg of lubricant type A every 6 months.
   - Worm gears, KV 60-20 LE
     - See Katsa maintenance instructions.
   - Gear motor, Bauer DP2A122CFG1-171/309MG
     - See Bauer maintenance instructions.
3. Inclination adjustment mechanism
   - Lifting screws
     - Use 0.05 kg of lubricant type A every 6 months.
   - Worm screws KV 100.2-37FK1
     - See Katsa maintenance instructions.
   - Universal shafts
     - Use 0.05 kg of lubricant type A every 6 months.
   - Gear motor Bauer DPA122CFG1-171/309MG
     - See Bauer maintenance instructions.

PMA-05/D-4
Rev. 03-25-91

14

---

help reinforce or clarify the content information (see Figure 11.A.2). All definition key words are in boldface type, followed by the respective definition. The respective key words in the content information are in italics so the reader will know that the specific word is a key word and that a definition is provided on the right third of the page. Also, simple illustrations with labels help the reader visualize the content information and the working of the respective component.

Finally, the document computer file name and revision date are provided

Figure 11.A.6 Sample Progress Review Page (Page 20) From the *Headbox Training Manual*
Developed by Alabama River Newsprint at Perdue Hill, AL

ALABAMA RIVER NEWSPRINT TRAINING MODULE

Headbox

PROGRESS REVIEW

The following Progress Review will test your knowledge of the information in this module. If you miss some on your first attempt, go over those items on which you are weak and try again. Again, this is for your own benefit, try to answer all of the questions before looking at the Answer Key on the following page.

1. The primary function of the headbox is to transform a _____ input flow into a stable _____ jet stream that flows at a continuous rate over the entire width of the machine.

2. The headbox is designed with a heating system to keep the temperature on both the inside walls and outside walls as close to the temperature of the _____ as possible.

3. All flow surfaces in the headbox have been _____ except for the holes in the perforated plates between the main sections to keep the headbox clean and keep pulp from sticking to the headbox walls.

4. The headbox receives accepted stock pumped by the _____ _____through the back side of the machine into a tapered inlet _____ then must reverse direction and flow through a tube bank.

5. A _____ _____ is fitted from the front to the back of the header for determining the pressure flow through the header.

6. The function of the _____ _____ is to discharge the stock in the headbox into the nip of the twin-wire former at a velocity close to the speed of the wire.

7. The SYM-FLO HS Headbox has _____ _____ _____ which feed stock from the inlet header to the turbulence generator and the slice section.

8. The top slice lip is fitted with _____ spaced 4.72 inches apart across its length to control the jet profile shot onto the wire of the paper machine.

9. The _____ _____ _____ cannot be started until the wire section and primary screen are running.

---

in the left side of the footer (see Figure 11.A.2), while the page number is shown in the far right-hand side of the footer directly in line with the end of the left margin line separating the header from the content information.

The pages illustrated in Figures 11.A.1 through 11.A.6 are excellent examples of a functional, effective, and an attractive document that is likely to attract the attention of the readers. This fact is supported by an excerpt from an article on training manuals that appeared in the December 1993 issue of *Training.* The excerpt is as follows:

Appearances Count — If a manual looks too technical, it will intimidate readers. Page layout goes a long way toward making information accessible and easy to understand. Make sure to include plenty of white space on each page. Break up blocks of text with bullets, subpoints and highlights, and number steps in procedures. On the other hand don't go overboard on design; each element should make the information easier to understand.(*Tips for Writing 1993, 15*)

Because of the importance that the training manual page layout plays in the reader's ability to comprehend the content, the layout must be "roughed out" prior to beginning the actual writing of the content information so that the writer knows how and when to add such things as headings and subheadings. The term roughed out is used because the employee who will be responsible for "placing" or "importing" the content information may take some artistic liberties, provided they are consistent with the layout so that all elements flow in an attractive format and aid reader comprehension. The only exception to giving artistic liberties to the employee responsible for importing the text into the desktop publishing document occurs when a selected format has already been used in other company manuals. In such instances, the technician must stick with the predetermined layout. Even then, there will be some artistic decisions made, such as where to place page breaks and illustrations.

5. **Writing of the Training Manual Content Information.** The writing of the training manual content information can be done in several ways. First, if the company or department is small, one technical professional may be assigned the writing responsibility, assuming that the individual knows the content information well enough to write about it and that he or she can write. Another way is to use a company technical writer who, in all probability, is a professional who is familiar with the technical content and can write. The final way is to take a company technical writer who may not be a technical professional but who is an experienced writer and who can grasp technical content information with the help of department engineers, vendor manuals, and other technical manuals.

Each of the ways to conduct technical writing has advantages and disadvantages. Although the best solution would be to use a company technical professional who can also write, most companies will not be able to spare such talent on writing training manuals even if that is the best approach. No matter who does the writing, the process will be the same, as illustrated in the flowchart in Figure 11.B. Large technical writing firms use a standard 30-step process, while the in-house process can effectively be boiled down to a practical 12-step process that takes into consideration all essential elements from research to quality assurance. In fact, as reflected in Figure 11.B, the flowchart is divided into five major groups: (a) Research, (b) Development, (c) Editing, (d) Validation, and (e) Final Editing.

Figure 11.B Training Manual Development Flowchart

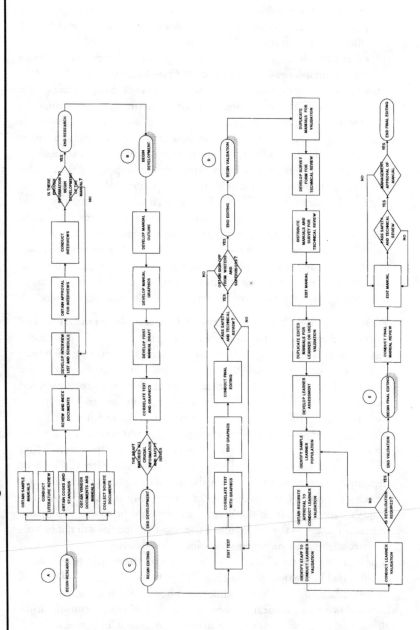

a. *Research*

Research is based in part on the information obtained from the initial project research stage described earlier. It is from this initial information that the training manual outline was developed. The next step is to conduct an in-depth review of all of the material collected during the initial research stage, including vendor technical manuals. The next step is to conduct on-site interviews with users of the respective equipment or process. If the equipment has been at the company site for some time, then the task will simply involve conducting interviews with the employees who work with the equipment on a day-to-day basis. If the equipment is new to your company, then you may consider having the equipment vendor make arrangements to conduct an on-site visit to a customer who has been using the equipment for some time and conduct the interviews with those employees.

The first step of the interview process is to develop an interview form. The interview form should include the name of the interviewee, job title, work location, and space for job summary and job task analyses associated with the operation of the equipment. The interview questions should be based on information gathered from the vendor manuals, technical manuals, and other training manuals. Careful attention should be given to developing step-by-step information (procedures) on the required operations used and verifying the inclusion and order of these steps with all employees and vendor technicians. Equally important, obtain input from equipment and process users on how the entire process can be separated into steps or procedures. As a minimum, at least two operators should be interviewed to verify the respective process. If only one operator is available, then interview the operator's immediate supervisor, who should have a working knowledge of the equipment to verify information gathered from the operator interview.

b. *Development*

The development stage consists primarily of writing the first draft of the content information with appropriate links with illustrations and key word or terminology references. This stage is appropriately described by Casady in an article titled "The Write Stuff for Training Manuals," as follows:

> Use your outline as a guide in drafting the first lesson, section, or unit. Material should be presented logically — generally beginning with relatively simple tasks and progressing to more complex ones — and in the most efficient sequence of steps.
>
> It is vital to make use of your own creativity rather than rely on another person's writing (such as the text of a supplier's

training manual). Copying someone's else's writing is illegal. Without proper attribution, even paraphrasing from another work may be illegal. Besides, copying or paraphrasing may incorporate another manual's weakness into the new work. As a writer, your goal should be to produce a better manual than any existing ones.

Break down every task described in the manual into manageable learning segments. Concise language with active verbs is best. Long sentences and unwieldy paragraphs may confuse users and cause them to miss important steps. Instead, use short phrases and numbered steps. And an outline format can help readers clearly identify subpoints of substeps.(Casady 1992, 18)

c. Editing

The editing will be completed by an individual who was not responsible for writing the first draft. The editor must have good writing skills, be well versed in common grammatical mistakes made by writers, and be committed to consistency and quality. As a minimum, the editor should follow the recommendations and information provided in Chapter 5. Besides common grammatical errors, the editor should review the first draft for clarity. Often, the editor has one of the most difficult jobs because he or she must be objective while still ensuring that the initial writer is not offended by the editing. Due to the many hours that are put into development of the draft, ownership and pride become well ingrained in the process; inexperienced writers can find it difficult to accept constructive criticism. Most writers will attest that having one's work edited by another can be both a stressful and humbling process that becomes more acceptable with experience and is an important phase of the technical writing process.

d. Validation

Validation is extremely critical to the entire training manual development process. Without validation, there is no certainty that the information and procedures provided in the manual are accurate or complete. The validation process incorporates obtaining as much constructive input on the accuracy and completeness of the manual as possible. The first step of the validation process incorporates giving the draft manual to current operators and vendor technicians who are familiar with the operation of the equipment. During the process, the reviewers should be given a form where they can make general comments and sign, showing that they have reviewed the draft. The accompanying instructions to the reviewers should incorporate a statement that they should verify that all operations, steps, and even theories are, to the best of their knowledge, accurate. If the information is not accurate, then the reviewer should provide corrections or changes that will correct any deficiencies.

Based on the reviewers' comments, corrections and changes should be made as needed. The draft then goes back to the editor for editing. The next step is the learner validation process, which consists of giving the training manual to a small sampling of learners who have the same characteristics as the learners who will eventually be using the manuals (experience, education, etc.). The learners should be given the manuals to read, along with a short quiz designed to test comprehension of the material covered in the manual. Also, another good step is to have the learners do a mock review or check-out on the actual operation of the equipment or processes covered in the manuals. Any deficiencies identified during this phase should be corrected and sent to the editor for review. If the deficiencies were significant, scoring below 90 percent on the tests and mock check-out, then the process must be revalidated after corrective actions and editing have been completed.

e. *Final Editing*
The final editing process consists of an exhaustive review of the document to ensure accuracy and consistency. Also, particular attention should be paid to proper grammar and spelling. After the final editing stage has been completed, the training manual is ready to print the camera-ready version required for printing, as discussed in Chapter 12 on the production of technical documents.

Final Activities

Once the final edition of the training manual has been completed, then it is ready for duplication (or printing) and distribution. Following the recommendations provided in Chapter 12 should help facilitate this important last step.

DOCUMENTATION, DUPLICATION, AND DISTRIBUTION OF THE TECHNICAL DOCUMENT

"In the past, especially in the last half century or so during which graphic design as a commercial art has flourished, people entered the field through formal training in art schools and apprenticeships with experienced designers. The almost overnight proliferation of desktop publishing technology has attracted and, through management expectations, forced many people with no training in the visual arts to take responsibility for a wide range of printed material. Increased access to publishing tools has motivated many businesses to produce in-house publications that were previously done, in whole or in part, by outside contractors. At the same time, the promise and the inevitable hype surrounding desktop publishing has raised expectations about internal and external communications of all kinds" (Shushan and Wright 1989, xi).

R. Shushan and D. Wright
Desktop Publishing by Design: Aldus PageMaker Edition

Introduction

Production of the technical document is a critical step that is usually not given much attention. Again, more often than not, individuals wait until the last minute before any consideration is given to those critical steps that must be properly conducted in order to produce a quality technical document. For purposes of this publication, production of the technical document incorporates four essential processes or steps: (1) Documentation, (2) Duplication of the Technical Document, (3) Quality Control, and (4) Distribution of the Technical Document and Proprietary Information. The remaining portion of this chapter includes important information and recommendations related to the production of the technical document.

Documentation

In many instances, the crucial element of proper documentation of sources and information is neglected for several reasons. First, many technical professionals wait to the last moment before even addressing the need to give credit where it is due. The result is that they misplace papers or simply forget that there are several undocumented references or citations. The next reason involves the darker side of human nature: a few technical professionals will take credit for the information — usually called plagiarism — instead of following established professional ethics and procedures associated with proper documentation. This practice has ruined the careers of many otherwise brilliant professionals because they claimed credit for the work of others or even presented conclusions or findings based on fabricated data or information.

The remaining portions of this section provide information and recommendations on three elements associated with the correct and ethical practices associated with documentation: (1) Plagiarism, (2) Footnotes and Endnotes, and (3) References and Bibliographies.

1. **Plagiarism.** Today, course work at both the secondary and higher education levels includes sections on what constitutes plagiarism and what should be done when writing a document to ensure that proper credit is given authors. Instruction is usually given in English courses. As professionals continue their education, proper documentation is similarly covered in greater detail through the requirement of theses and dissertations. Additionally, many business engineering, and medical schools provide required course work in ethics, which always includes sections on ethics in technical writing.

   Even with all of this education and training, there is a small number of professionals who find that they have been embarrassed or, worse yet, that their careers have come to a screeching halt because they did not give proper credit or were not ethical in their responsibilities in writing a technical document. By following three simple rules, these situations could have been avoided and still resulted in a high-quality technical document. These rules are as follows:

   a. Document all sources of information properly. This documentation should follow the standards associated with the style manual being used. This documentation includes giving credit for long citations as well as quotes of only a few words.

   b. To ensure proper credit is given on all sources and quotes, add the documentation during the writing stage of the technical document. It is almost impossible for writers to remember where specific quotes originated, so if they wait until the technical document is completed, they will forget to provide documentation for some sources and quotes.

   c. Do not change or modify information or quotes. If clarification needs to be provided, follow the associated rules in the style manual being used. Usually brackets can be used to add clarification. Brackets should be used only in extreme instances. If too much clarification is needed, then consider using another quote or other information.

In addition to the foregoing rules on giving credit where credit is due, there are four others associated with ethics in technical writing as developed by Robert Goldbort. These rules are as follows (Golbort 1993, 52):

a. Cite only those publications pertinent to the research reported and that provide background work for the reader.
b. Avoid using excessive journal space and complicating literature searches by fragmenting the research papers derived from an extensive study.
c. Unless you have obtained explicit permission from the initial investigator, do not use information obtained privately in conversation or correspondence, or when reviewing manuscripts or grant applications.
d. Restrict criticism of a study to the research; any personal criticism is irrelevant and invidious.

2. **Footnotes and Endnotes.** Footnotes are placed at the bottom of a page to provide additional information or clarification of information given in the text of the respective page or to provide a full bibliographical reference of the cited information or quote. In turn, endnotes contain the same information as footnotes but are placed at the end of a chapter or document. The rules of the style manual being followed should determine if footnotes or endnotes should be used. This publication uses author-date-type citation, where the author and date of the cited material are included at the end of the respective citation or quote. The full bibliographic record for the article or publication is presented at the end of the publication.

Beyond those citations where footnotes are used to provide bibliographic information, the actual use of footnotes or endnotes (hereafter called notes), however, should be given serious consideration. The overuse of notes is not only distracting, but can be extremely threatening to many readers. Thus, notes should be used only when necessary to help clarify information in the text or to direct the reader to additional information that would not be proper to place in the text. If a technical professional finds that there are too many notes in a document, which will vary depending on length and technical content, then a determination must be made as to whether the document is properly written. If too many notes have been placed in the document, then perhaps the writer did not do a good job in presenting the textual material in proper detail.

Unfortunately, today's high-end word processors make it too easy to use footnotes. With the touch of a computer key, a note is numbered and automatically inserted in the document. The ease of developing notes has helped lead to overuse. Years before the arrival of the word processor, writers avoided the use of notes, especially footnotes, because of the difficulty in ensuring that a note was properly numbered and placed on the same page as the related textual material. Now the opposite is true, and notes are overused, making the technical document look cluttered and unprofessional.

If notes must be used (a personal decision that the technical professional must make), consider using endnotes. Endnotes will not be as distracting to the reader as footnotes and still provide the reader with the availability of the additional information if he or she is interested in flipping to the end of the chapter or document.

3. **References and Bibliographies.** A full citation of a reference must be provided for all sources of information or quotes used in a technical document when the author-date–type of documentation is used. The style manual used will provide details on placement and kinds of information to be included in a reference list, which is usually placed at the end of the document. It is important to provide sufficient information so that the reader can obtain a copy of the referenced publication with little effort. Finally, the reference list is not a list that one uses to impress the reader. In other words, only list references if you have cited actual quotes or information from them in the text of the technical document.

Similarly, bibliographies include essentially the same information as the reference list and should contain only references for publications or information cited in the text. A standard bibliography does not include lists of books read during development of the technical document not cited in the document. Such lists can be used separately (included as an appendix) as "suggested reading," according to the style manual being used.

Duplication

How the technical document is copied or duplicated for distribution is as critical as the layout of the document. As discussed in Chapter 5, the duplication process is an extremely critical factor when graphics are used because of color and gray scale considerations. If graphics with colors or gray scales are not used, then how the document is duplicated is not as critical a factor.

In the case of the technical document that does not contain color, color graphics, or gray scales, the document usually can be duplicated by the standard copy machine. Today's copy machines are sophisticated and can handle duplex copying (copying both sides of a page simultaneously) and can even staple and collate the documents. If the number of copies is small (under 100 copies) and you have access to a copy machine that produces quality copies, then there is no reason why the copying cannot be conducted internally. On the other hand, if the number of copies is more than 100 and the document is large, then it may be best to send the work to an outside copy shop.

If you plan to copy the document internally, there are several steps that should be taken to ensure that your copied document is of a high quality—a clear and crisp copy without any random black marks or smudges. These steps are as follows:

1. Several days before the document is ready for copying, conduct a test run of the copy machine to make sure it is working properly and that copies are of a high quality. If there is even the hint of poor quality or equipment failure, contact the authorized service representative to conduct repairs or preventive maintenance on the copy machine.

2. Have a second plan in the event that the copy equipment fails. If you are faced with a deadline, such as submitting a proposal, not having a second plan can mean that your proposal may never be copied and submitted; such a situation could be hazardous to your career. The second option could be taking the document to an outside vendor for copying. If you are located in a rural or isolated area, the backup plan could even be taking the document to another noncompeting area company that has sufficient copying capability. Plan for unforeseen duplication problems such as equipment failure.

3. The original document should be printed on a laser printer using high-quality paper made especially for laser printers. True, such paper costs a few more cents per page than regular duplication paper, but the print quality is well worth a few pennies. The print on laser printer paper produces crisper and cleaner copies with the purest black tones. A high-quality master will produce a higher quality copy, even though the actual copies may be printed on regular duplication paper. If the document being duplicated is a proposal or other item to be distributed outside your company, it would also be a good idea to purchase high-quality duplication paper. When purchasing any type of duplication paper, it is important to keep in mind the texture of the paper and how the dry ink will interact with the paper. Again, conducting a test batch several days before the final version will be needed makes certain you will be getting the desired look and quality.

4. The type of binding used for the document is also important. It is best to "keep it simple." One could spend a great deal of money on special bindings with custom logos, but one problem is that most of these bindings simply do not last. More often than not, bindings that rely on glue will, in time, come unglued. For internal documents, using simple three-ring binders with customized cover inserts is not only practical but extremely functional. The binders can take a beating without any damage to the document pages, and they are specifically designed to be opened and laid flat on a table surface, which makes it easier to read and to make notes in the document.

Another durable binding format is the plastic comb binding. They are somewhat cheaper than three-ring binders and are functional, attractive, and easy to assemble provided the appropriate binding equipment is available. The comb-type assembly equipment can be purchased from almost any office-supply store starting as low as several hundred dollars. The plastic combs cost only a few cents each, with the cost increasing as the thickness of the document increases. As with the three-ring binders, customized binder covers can be used for only a few additional cents.

5. Although copying on both sides of a page saves money, the results are generally not worth the few pennies saved, so take a few extra cents and copy on only one side. This is especially important if you are forced to use low-quality paper for duplication; the printing on one side of the document can be seen on the flip page.

6. If the document is large (over 10 pages), a table of contents should be provided along with index tabs for chapter or section dividers to help the readers find important sections. Avery produces laser-printer index tab kits of varying cuts (number of tabs). These kits are easy to use with any word-processing program and laser printer. In fact, Microsoft Word has built-in templates for almost all of the Avery kits.

7. Before you start the actual copying, ensure that you have extra copy toner and any other supplies that your specific machine requires. Clean the glass on the copy bed with a manufacturer-approved solvent to ensure there are no marks or streaks on the copies. Often black spots appear on copies, usually caused by dried correction fluid left on the glass from previous copies made while the fluid was still wet.

If the technical document contains color, color graphics, or gray scales, it should be printed by a professional printer to whom you provide a camera-ready document. The actual printing process varies depending on the type of graphics in the documents and whether color separations are required. Using an external printer is more cost-effective than most people think — especially when you consider the labor time involved in printing as well as the cost of printing materials and supplies. If the print job is over 100 copies, the final cost of using a professional print shop is actually cheaper than producing the documents internally. Furthermore, the professional printer will be able to provide a higher quality print of graphics in the technical document.

Depending on the type of document and the intended audience, the printer can help you decide the best printing process to use, as well as the best materials. A professional printer will be well versed in the interaction of colors and other considerations; ask for his or her advice during the development of the technical document. Taking this into consideration, there are five recommendations that should be followed when using the services of a professional printer for the duplication of the technical document. These recommendations are as follows:

1. Plan ahead and select your printer well in advance of the scheduled print date. It is strongly recommended that you visit the printer during

the document development stage to get input on formats, colors, and gray scales that will be used.

2. If you are not dealing with extremely confidential or proprietary information, provide a copy of the document draft so that a test print can be conducted. Depending on the type of printing process, a test run may be cost-prohibitive. If the job is extremely important, the printer may be willing to conduct a test run for a few extra dollars. Of course the smaller print shop will generally be more receptive to conducting such test runs than the bigger operations where the printing equipment is being run 24 hours a day. This test run of the draft will provide you with an excellent opportunity to see how the ink and paper interact, as well as an opportunity to see the print quality of the document graphics.

3. Inform the printer in advance that you expect the highest quality possible and that you will not accept or pay for poor-quality copies.

4. Ask to see a set of proofs before the whole job is run. Again, depending on the type of printing process used, this step may mean that you will have to be at the printing site just before the scheduled print run. When reviewing the proof, do a final check to ensure that the correct version of the document is being printed and that all corrections or changes have in fact been incorporated. Although you may have originally submitted a camera-ready document, a few last-minute changes are usually made by the printer at the request of the customer.

5. Keep in direct contact with the printer as the document is being completed. Make regular telephone calls to the printer to make sure that there are no miscommunications on the scheduled print date.

Quality Control

The last step before distribution of the technical document is to conduct a final quality-control review of the copies to be distributed. The premise behind a quality-control review is that it should not be assumed that all copies are intact and contain all of the pages, or that all of the copies are printed properly. There are at least five quality-control review steps that should be conducted:

1. Assign the responsibility of the quality-control review to an experienced and responsible staff member, who ensures that all quality-control review checks are conducted.

2. Review each document to ensure that all pages are included and are in the proper order. Usually one or more pages are missing in one or more documents. Also, this person should ensure that quality of print is good for each document page, and the pages are properly aligned.

3. Check bindings and binders for defects. Any bindings or binders that are not properly constructed or are damaged should be rejected.

4. Check document covers for quality, which includes ensuring that the cover-page inks cannot be smudged or removed under normal use conditions.
5. Inspect index tabs, if used, to ensure that they are properly and securely fastened or bound and are properly located in the document.

Distribution of the Technical Document and Proprietary Information

The distribution of a technical document is the final step of the document development process. Many technical professionals do not associate the distribution process with the development process. However, if the technical document is never received by the intended reader, all the work that has gone into its development will be wasted. The technical professional should not take for granted that the document has been distributed as instructed. In today's workplace, job responsibilities are fragmented and most employees are overstressed and overworked. Also, a technical professional cannot assume that his or her instructions to a subordinate will be acted upon immediately — this is especially true when such employees report to several supervisors. Extra care must be given to how technical documents that include proprietary information are distributed and stored before the distribution process begins. Proprietary information includes proposals, marketing plans, production plans and reports, as well as any other type of information that is of strategic importance and must be kept secure from competitors. Because of the very nature of technical documents, it is best to treat them as documents that contain proprietary information.

There are seven steps that should be followed to ensure that the completed technical document is distributed as intended:

1. Develop a transmittal list of all recipients who have been approved to receive a copy of the technical document. This list should include name; title; company, organization, or department; and complete mailing address, including mail stop.
2. Place the document in a security envelope and label with all of the information specified in step 1. A security envelope contains a special seal that indicates if the envelope has been opened. A security envelope is shown in Figures 12.A and 12.B, which demonstrate how the recipient can tell if the envelope has been opened.
3. Specify how the document is to be transmitted to each recipient. For internal recipients, it is best to have a responsible employee personally deliver the document to ensure that the recipient has received the document and that it has not been unattended around an office or mail bin.
4. Attach a letter of transmittal to each document. Also, internal recipients should be required to sign the transmittal form, showing that they have received the document.
5. Send the technical document to all external recipients via a responsible

Figure 12.A Picture of an Unopened Seal Window of a Security Envelope Produced by the International Envelope Company

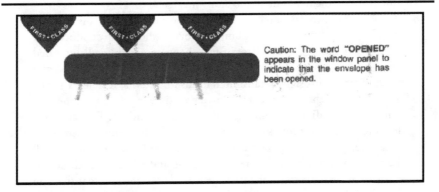

Caution: The word "OPENED" appears in the window panel to indicate that the envelope has been opened.

Figure 12.B Picture of an Opened Seal Window of a Security Envelope Produced by the International Envelope Company

Caution: The word "OPENED" appears in the window panel to indicate that the envelope has been opened.

courier service such as Federal Express. The courier service should be asked to pick up the package; you should not have the package dropped off at a delivery point. By having the courier pick up the document, you will receive a signed receipt showing that it was handled by a courier employee. Such proof may be extremely critical in the event that an important document such as a proposal was never received by the intended person. In a few instances, if a company can show proper documentation that the document was in fact picked up by the courier service but never delivered, most customers will allow a company to resubmit their bid.

Because of this documentation of transmittal and receipt, you will have a paper trail showing chain of custody. Couriers such as Federal Express and United Parcel Service (UPS) also provide free computer software that can be installed on almost any modem-equipped computer to provide users with a full spectrum of services such as requesting a pickup to actually tracking the document and obtaining details on the time the document was delivered and who signed for it.

6. No more than two days after the activities associated with the distribution of the technical document, a complete review should be conducted to ensure that the document has, in fact, been received by all intended persons. This process incorporates taking the initial distribution list and matching it against signed transmittal forms for internal recipients as well as a printout of the computer tracking report showing the date, time, and name of the individual who signed for the document. For technical documents such as proposals that are required to be in the possession of certain recipients by a certain date and time, it is a good practice to call the recipient a few hours before the deadline and ask if the package was received. In the rare event the package was not received by that time, an alternate plan would be to send over another document immediately if it is in the same vicinity or even consider faxing it. Today, many companies will accept a faxed proposal if the sender can show that a good-faith effort was made to get the document to the designated location on time (documented by a receipt showing that the courier had the document well before the required deadline). The main point to remember is that the entire document development job is not complete until all of the recipients, especially the main recipient in time- and date-sensitive proposals, have in fact received the document. Only in extreme emergencies, if a proposal was lost in transit by the courier services or other extenuating circumstance, should consideration be given to faxing a proposal to a customer.

7. Copies of the document should be filed or stored according to the company policies and procedures. All computer files associated with the technical document should be backed up and stored. Finally, the computer files on the computer should be deleted if they are on a computer that is not secure.

THE TECHNICAL
PROFESSIONAL'S
PERSONAL LIBRARY

One of the most important things a technical professional can do to help enhance his or her career is to establish a personal technical document library. The collection activities should start in college and continue throughout a career until retirement. Such an undertaking will need a great deal of planning, hard work, and strong persuasive skills to convince other technical professionals that they should part with a copy of a document.

Introduction

Almost every technical professional will be required to develop many technical documents throughout his or her career. Sometimes these will be routine documents that can be developed based on accepted company standards and practices. In other instances these will be documents that are special in nature and deal with a subject or problem that cannot be adequately addressed using technical document formats adopted by the company. Such situations put added stress on the technical professional to develop a document that is professional looking and includes all the elements required to adequately address the problem or project.

The remaining portions of this chapter provide suggestions and recommendations on how a technical professional could start and maintain his or her own technical document library. Having a comprehensive library will provide the professional with a wealth of technical document formats that will be a source of inspiration during those critical moments of searching for ideas and formats to match the current technical writing assignment. The technical professional also needs access to new telecommunications technologies to initiate contact with other professionals who can provide an excellent support network of helpful ideas and suggestions during both the research and writing stages. The remaining portion of this chapter is divided into two sections: (1) Collecting Technical Documents for Future Reference; and (2) Computer Forums.

Collecting Technical Documents for Future Reference

One of the most important things a technical professional can do to help enhance his or her career is to establish a personal technical document library. The collection activities should start in college and continue throughout a career until retirement. Such an undertaking will need a great deal of planning, hard work, and strong persuasive skills to convince other technical professionals that they should part with a copy of a document.

Having a comprehensive technical document library will prove most helpful during both the research and development stages of a document. By reviewing the collection of technical documents, the technical professional will be able to get some ideas on both the format and content of the current project. This does not imply that the technical professional would copy any information from another technical document without giving credit to the original author, but rather that the documents in the library will help inspire creativity in the professional.

Prior to starting your own technical document library, a couple of issues must be addressed. First, select the location of your library. The initial response of most dedicated technical professionals would be to keep such documents in their offices. Keep in mind, however, that if the documents are maintained in the office and have been collected during normal work hours, then the company could reasonably say that your library is actually company property. If you leave the company you may not be allowed to take the library with you. It would be beneficial to get written authorization to maintain a personal library in your office with the explicit understanding that the library is yours to take if you ever leave the company. If you cannot get such authorization or feel uncomfortable about asking for it, then the next best step would be to maintain the library at home.

Regardless of where the technical document library is housed, the information collected and maintained will be the same. Generally, there are many sources from which you can collect sample documents:

1. Technical documents that you have written.
2. Technical documents that have been sent to you by others (both solicited and unsolicited).
3. Technical documents received from other companies that you have specifically requested for your library.
4. Technical documents that you have received from colleagues at other companies.
5. Technical documents from public-domain projects, where a specification request can be made under the provisions of the *Freedom of Information Act (FOIA)*. Under the provisions of FOIA, copies of most federal records (excluding records related to national security, government personnel files, criminal investigations, etc.) can be obtained through a written request, and the agency must provide it within 10 working days of the request.

All of these are viable sources and should be actively pursued. In fact, the technical professional must be persistent in collecting technical document samples. You will quickly realize that many professionals tend to be a little bit "stingy" with their documents and will not always cooperate with such requests. The best way to overcome such resistance is to agree to trade it for a technical document of equal value from your collection. In essence, the process is similar to the way kids, and some adults, trade professional sports cards.

In collecting company documents for your own personal technical document library, keep in mind that your company may have a policy or work rule against taking such documents. Again, before taking any documents, or copies of documents, away from the office, obtain written authorization from your immediate supervisor.

Finally, a comprehensive filing system of the technical documents should be developed. Such a practice may seem silly when you first start your library collection, but over time such a system will be a necessity unless you want to spend hours looking for a specific technical document. Ideally, the filing system would group documents according to type of technical document, with a cross-reference to technical document subject. Following such a format, you should be able to find a technical document related to almost any subject and document type within a matter of minutes. Such a system could be easily maintained in a filing cabinet in a cool dry place so that the technical documents do not deteriorate significantly over time. Putting the cabinets in a garage subject to extreme hot and cold temperatures or in a basement that is usually quite humid can reduce the life of most technical documents from 20 to 30 years to under 10 years. Also, the files should be in a location where bugs, mice, or other pests cannot damage them.

Computer Forums

Several of the computer on-line and Internet services such as CompuServe have special-interest groups called forums that are an excellent place to network and make professional contacts. As discussed in Chapter 2, these forums are an excellent source for conducting research and asking other technical professionals questions about projects you are working on. For an example, you may post a question on the forum about the latest research on an engineering process such as geometric diminishing and tolerancing. Within a few hours you will receive messages from others who are working on similar projects or who know other professionals you could call.

Besides the e-mail contacts with other professionals, however, there are other equally important resources available on these forums. For example, in the Engineering Automation Forum (called LEAP) on CompuServe, there is an electronic library with all types of valuable technical documents that can be downloaded. Many technical professionals are savvy enough to realize how valuable such resources can be when they are working on professionally important technical documents. Figure 13.A, for example, shows the menu of the library sections of the Engineering Automation Forum

Figure 13.A LEAP Forum Topic Area

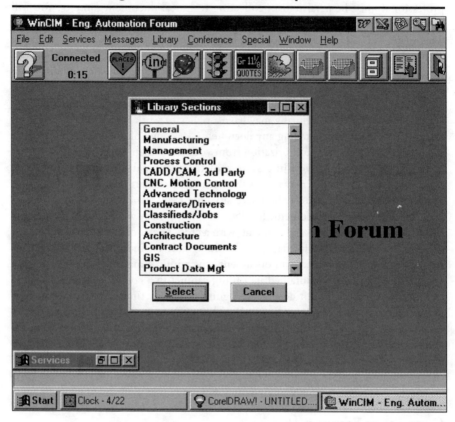

on CompuServe. Figure 13.B includes a list of document files that can be downloaded from this forum.

Because of the almost instant availability and low cost of obtaining these documents, it is a good idea to "cruise" these forums on a regular basis and download files that may be of interest to you or related to your job responsibilities. Naturally, as these files are posted for public use, you will be able to freely use these documents or the information in any technical documents you write. As with all information taken from other sources, you must credit the source.

Most current style manuals have a reference format for information that was obtained from electronic bulletin boards and on-line services. For example, a definition of "research libraries" used in Chapter 2 was obtained from the CompuServe reference section, which contained the *1994 Information Please Almanac*. The specific on-line reference for this citation is as follows: "Research Libraries," in *The 1994 Information Please Almanac* [database on-line] (New York: Houghton Mifflin [cited 12 February 1995]). Available from CompuServe, Columbus, Ohio.

Figure 13.B Sampling of Process Control-Related Documents Available in the LEAP Forum

Title	Size	Date	Accesses
LoopMate Loop Tuning Software	1885922	3/25/96	1
Control Draw Demo information in	16058	1/31/96	71
Control Draw Disk 1 of 2	1368678	1/31/96	17
About HydCalc - Hydraulic Calcula	1066	1/31/96	11
Control Draw disk 2 of 2	1269269	1/31/96	16
ACDSIZER	115512	1/31/96	6
PMAdvisor v1.0	120263	1/31/96	16
PRISM 9000(c) for Windows(tm)	816248	12/30/95	19
TV MEASURMENT GENOP ISRAE	251392	11/22/95	4
Steam properties v1.0<ASP>	526743	11/14/95	49
Data acquisition using your compu	119062	11/9/95	106
steam tables asme	61568	10/31/95	75
VP: High Res I/O VBX to 5 mS	45779	10/18/95	70
FORTH LANGUAGE MS-DOS	334240	10/18/95	33
Scorpion Password	52	10/13/95	30
Scorpion Job Processing System	1012494	10/3/95	18
INSTINDX V1.1 file 2 of 3 require	350000	9/25/95	19
INSTINDX V1.1 file 3 of 3 reg for	350000	9/21/95	22
INSTRINDX V1.1 1 OR 3 REQ FII	496434	9/21/95	21
Varible Value Reader for Bailey D	39194	9/5/95	22
Press Release for Modcomp/Goul	5206	8/17/95	16
Queue Freedom Bridge Press Rele	4776	8/17/95	3
Measure-Up for Windows Press P	3754	8/9/95	28

Description   Mark   Retrieve   View   Close

REFERENCES

Accreditation Commission Accreditation Board for Engineering and Technology, Inc. (ABET). 1996. *Criteria for Accrediting Programs in Engineering for Programs Evaluated During the 1996-1997 Accreditation Cycle* [online] [cited 20 December 1996]. Baltimore, MD: Engineering Accreditation Commission Accreditation Board for Engineering and Technology, Inc. Available from World Wide Web: http://www.abet.ba.md.us/EAC/96eac.html.

Advanced Accessory Systems. 1995a. *Standard Operating Procedure: Preventative Maintenance.* Shelby Township, MI: Advanced Accessory Systems.

Advanced Accessory Systems. 1995b. *Work Instructions Manual: Statistical Techniques.* Shelby Township, MI: Advanced Accessory Systems.

Alabama River Newsprint Company. 1993a. *Alabama River Newsprint Training Manual: Chemical Preparation (ARR–03).* Perdue Hill, AL: Alabama River Newsprint.

Alabama River Newsprint Company. 1993b. *Alabama River Newsprint Training Manual: Headbox (PMA-05).* Perdue Hill, AL: Alabama River Newsprint Company.

Amendments to the Telephone Consumer Protection Act of 1991. Fed. Reg. 57,206 (1992) (to be codified at 47 CFR § 68.318(c)(3):48336).

Anthes, G. H. 1995. Info highway hits technical detours. *ComputerWorld* (February 6): 20.

Bennett, A. W. , D.McAuliff, E.B.Makram, and A.A. Girgis. 1990. Introducing effective technical communication (ETC) in power engineering curriculum. *IEEE Transactions on Power Systems* 5, 4 (November): 1384–387.

Brandt, J. 1995. Research tools in educational research forum. CompuServe electronic mail (28 January 1995).

Brisbin, S. 1995. ISDN: The promise and the pitfalls. *MacUser* [online] 11, 5 (May) [cited 10 February 1995]: 105. Available from Alabama Southern Community College, Monroeville, AL.

Casady, M. J. 1992. The write stuff for training manuals. *Training & Development* 46, no. 3 (March): 17-23.

Cleland, D. I., and H. Kerzner. 1985. *A Project Management Dictionary of Terms*. New York: Van Nostrand Reinhold.

Connelly, J. 1995. How to choose your next career. *Fortune* [online] 131, 2 (February) [cited 10 February 1995]: 145. Available from Alabama Southern Community College, Monroeville, AL

Coopersmith, J. 1993. Facsimile's false starts. *IEEE Spectrum* 30, 2 (February): 46-49.

Copyright Office, Library of Congress. 1994, *Circular 1: Copyright Basics*. Washington, DC: U.S. Government Printing Office.

Corel Corporation. 1994. *CorelDRAW™ 5 - Volume 1*. Ottawa, Canada.

CompuServe. 1995. *Business Reference Brochure*. [Faxed from CompuServe on February 10, 1995]. Columbus, OH: Author.

Doe, J. 1994. Telephone interview with author, November 1994.

Dolle, R. 1990. Designing first impressions. *Journal of Environmental Health*. 53, 2:58.

Dollsion, J. 1994. *Pope-Pouri*. New York: Simon & Schuster.

Eklundh, K. S., and Carl Sjöholm. 1991. Writing with a computer: A longitudinal study of writers of technical documents. *International Journal of Man-Machine Studies* 35:723-49.

Elmer-Dewitt, P. 1993. Who's reading your screen? *Time* [online]. 141, 3 (January 18) [cited 10 February 1995]: 46. Available from Alabama Southern Community College, Monroeville, AL.

ERIC. 1992. *ERIC Processing Manual*. Rockville, MD: ERIC.

Feds may archive your e-mail. 1994. *Science News* [online] 145, 16 (April 16) [cited 10 February 1995]: 255. Available from Alabama Southern Community College, Monroeville, AL.

Fryer, B. 1994. Managing e-mail overload: Real problems, real solutions. *PC World* [online] 12, 4 (April) [cited 10 February 1995]: 43. Available from Alabama Southern Community College, Monroeville, AL

Goldbort, R. 1992. Ethics in scientific writing. *Journal of Environmental Health* 55, 2 (September/October): 52–53.

Good, C. V. 1973. *Dictionary of Education*. New York: McGraw-Hill.

Hart, J. 1993. Writing to be read. *Editor & Publisher* [online] 126, 45 (November 6) [cited 10 February 1995]: 5. Available from Alabama Southern Community College, Monroeville, AL.

Hartley, T. T., C. F. Lorenzo, and H. K. Qammar. 1996. *Chaos in a Fractional Order Chua System*. Linthicum Heights, MD: National Aeronautics and Space Administration. Report, NASA Technical Paper 35423.

Herbert, S. 1994. Progress reports. *Engineering* (January): 34.

Huntress, E. 1995 Technical writing in engineering automation forum. CompuServe, electronic mail (29 January 1995).

International Organization for Standardization (ISO). 1978. *Documentation – Numbering of Divisions and Subdivisions in Written Documents (ISO 2145)*. Geneva, Switzerland: International Organization for Standardization.

International Organization for Standardization (ISO). 1982. *Documentation – Presentation of Scientific and Technical Reports (ISO 5966)*. Geneva, Switzerland: International Organization for Standardization.

International Organization for Standardization (ISO). 1996. *Introduction to ISO*. Downloaded from http://www.iso.ch/infoe/intro.html on 31 May 1996.

Johnson, J. L., R. Insley, J. Motwani, and I. Zbib. 1993. Writing performance and moral reasoning in business education. *Journal of Business Ethics* 12:397–406.

Joseph, A. M. 1989. Wanted: Engineers who know how to write. *Tooling & Production*. 55, no.8 (November): 138.

Knight-Ridder Information, Inc. 1991. *Searching DIALOG: The Tutorial Guide to Commands*. Palo Alto, CA: Knight–Ridder Information Services, Inc.

Knight-Ridder Information, Inc. 1995. *Database Catalogue 1995*. Mountain View, CA: Knight–Ridder Information Services, Inc..

Krol, E. 1992. *The Whole Internet: User's Guide & Catalog*. Sebastopol, CA: O'Reilly & Associates, Inc..

Maki, P., and C. Schilling. 1987. *Writing in Organizations: Purposes, Strategies, & Processes.* New York: McGraw-Hill.

Editors of *Communications Briefings. Power Writing.* 1990. Alexandria, VA: Communication Briefings. Videotape.

Editors of *Communications Briefings. Mastering Memos.* 1992. Alexandria, VA: Communications Briefings. Videotape.

Microsoft Corporation. 1991. *User's Guide: Microsoft Equation Editor.* Redmond, WA.

Microsoft Corporation. 1993-1994. *User's Guide: Microsoft Excel — Version 5.0.* Redmond, WA.

National Information Standards Organization (NISO). 1995a. *Scientific and Technical Reports — Elements, Organization, and Design (ANSI/NISO Z.39.19-1995).* Bethesda: NISO Press.

National Information Standards Organization (NISO). 1995b. *Scientific and Technical Reports — Elements, Origination, and Design.* Bethesda: NISO Press.

Olsen, L. A., and T. N. Huckin. 1991. *Technical Writing and Professional Communications.* New York: McGraw-Hill.

Petroski, H. 1993. Engineers as writers. *American Scientist* (September-October): 419–23.

Pfaffenberger, B. 1992. *Que's Computer User's Dictionary.* 3rd. Ed. Carmel, IN: Que Corporation.

Ramsey, R. D. 1993. Improving your technical writing. *Supervision* 54 no. 7(July): 3.

Reference Software International. 1992. *Grammatik™ 5 User's Guide.* San Francisco, CA.

Reuter, F. 1994. The synthetic mind clashes with the reductionist text. *Skeptical Inquirer* [online]. 18, no. 4 (Summer) [cited 10 February 1995]: 404. Available from Alabama Southern Community College, Monroeville, Alabama.

Roberts, J. C., R. H. Fletcher, and S. W. Fletcher. 1994. Effects of peer review and editing on the readability of articles published in annals of internal medicine. *JAMA* [online]. 272, 2 (July 13) [cited 10 February 1995]: 119. Available from Alabama Southern Community College, Monroeville, AL.

Rudland, D.L., P. M. Scott, and G. M. Wilkowski. 1996. *The Effect of Cyclic and Dynamic Loads on Carbon Steel Pipe.* Washington, DC: U.S. Nuclear Regulatory Commission. Report, NUREG/CR-6438; BMI-2188.

Rudnikc, M., S. Wiener, and J. Kaplowitz. 1995. *Employee Communications and Technology: The_revolution.begins@now.* Stamford, CT: Cognitive Communications, Inc.

Saltzman, E. J. 1995. *Selected Examples of NACA/NASA Supersonic Flight Research.* Edwards, CA: National Aeronautics and Space Administration. Report, NASA Special Publication 513.

Samuelson, P. 1995. Copyright and digital libraries. *Communications of the ACM* [online]. 38,4 (April) [cited 10 February 1995]:15. Available from Alabama Southern Community College, Monroeville, AL.

Schriver, K. A. 1993. Quality in document design: Issues and controversies. *Technical Communication* (Second Quarter): 239-57.

Shieh, J., and R. A-L Ballard. 1994. E-mail privacy. *Educom Review* 29, 2 (March/April): 59-61.

Shushan, R., and D. Wright. 1989. *Desktop Publishing by Design: Aldus PageMaker Edition.* Redmond, WA: Microsoft Press.

Stewart, R. D. 1982. *Cost Estimating.* New York: John Wiley & Sons.

Stewart, R. D., and A. L. Stewart. 1984. *Proposal Preparation.* New York: John Wiley & Sons.

Strunk, W. I., and E. B. White. 1972. *The Elements of Style.* New York: Macmillan Company.

TABE Report for Forms 7 & 8. 1996. New York: McGraw-Hill.

The 1994 Information Please Almanac. Research Libraries. 1995.[database on-line]. (New York: Houghton Mifflin [cited 12 February 1995]). Available from CompuServe, Columbus, OH.

Thielsch, H. 1996. Letter to author, 31 March 1996.

Tierney, P. 1995. Message from the president. *Chronology* 23, 1 (January): 95:3.

Tips for writing a readable manual. 1993. *Training* (December): 14-15.

UMI. 1994. *Document Delivery Catalog.* Ann Arbor, MI.

Van Gorder, B., and N. Carter. 1995. Is the whole world going on-line? *American Visions* [online] 10, 2 (April-May) [cited 10 February 1995]: 38. Available from Alabama Southern Community College, Monroeville, AL.

Vanston, L. K., W. J. Kennedy, and S. El-Bardy-Nance. 1991. *A Facsimile of the Future: Forecasts of Fax Markets and Technologies.* Austin: Technology Futures, Inc.

Wellemeyer, C. G., S. L. Taylor, G. Jaross, M. T. DeLand, C. J. Seftor, G. Labow, T. J. Swissler, and R. P. Cebual. 1996. *Final Report on Nimbus-7 TOMS Version 7 Calibration.* Linthicum Heights, MD: National Aeronautics and Space Administration. Report, NASA Contractor Report 4717.

White, J. 1993. *On Graphics: Tips for Editors.* Chicago: Regan Communications, Inc.

Watterson, B. 1996. *There's Treasure Everywhere.* Kansas City, KS: Andrews and McMeel.

The World Book Encyclopedia 1995. CD-ROM on-line at Alabama Southern College, Monroeville, AL. New York: World Book, Inc.

SOURCE NOTES FOR FIGURES

Chapter 2

Fig. 2.A
Source: Screen shot reprinted by permission of the publisher, from *The Columbia Library System*® (database on-line). McGraw-Hill Library Systems (cited January 27,1996), Monterey, CA. Reproduced with permission of the McGraw-Hill Companies, Inc. Available from Alabama Southern Community College, Monroeville, AL.

Fig 2.B
Source: Screen shot reprinted by permission of the publisher, from *The Columbia Library System*® (database on-line). McGraw-Hill Library Systems (cited January 27,1996), Monterey, CA. Reproduced with permission of the McGraw-Hill Companies, Inc. Available from Alabama Southern Community College, Monroeville, AL.

Fig. 2.C
Source: Screen shot reprinted by permission of the publisher, from *Academic ASAP*™ (database on-line). 1996 Information Access Company (cited April 1, 1996) Foster City, CA. Available from Alabama Southern Community College, Monroeville, AL.

Fig. 2.D
Source: Screen shot reprinted by permission of the publisher, from *Academic ASAP*™ (database on-line). 1996 Information Access Company (cited April 1, 1996) Foster City, CA. Available from Alabama Southern Community College, Monroeville, AL.

Fig 2.E
Source: Screen shot reprinted by permission of the publisher, from *Academic ASAP*™ (database on-line). 1996 Information Access Company (cited April 1, 1996) Foster City, CA. Available from Alabama Southern Community College, Monroeville, AL.

Fig 2.F.1
Source: Reprinted by permission of the publisher, from "Cooling Bubbles Dissipate Heat," *USA Today (Magazine)* (serial-online). June 1995, 125, (n2601) (cited July 1, 1996): 1. Available from Alabama Southern Community College, Monroeville, AL.

Fig2.F.2
Source: Reprinted by permission of the publisher, from "Cooling Bubbles Dissipate Heat," *USA Today (Magazine)* (serial-online). June 1995, 125, (n2601) (cited July 1, 1996): 1. Available from Alabama Southern Community College, Monroeville, AL.

Fig 2.G
Source: Screen shot reprinted by permission of the publisher, from *University of Illinois at Urbana-Champaign Gopher Site* (database online). University of Illinois at Urbana-Champaign, Urbana, IL, updated January 26 1996 (cited February 5, 1996). Available from: Gopher site Gopher.uiuc.edu.

Fig 2.H
Source: Screen shot reprinted by permission of the publisher, from *University of Illinois at Urbana-Champaign Gopher Site* (database online). University of Illinois at Urbana-Champaign, Urbana, IL, updated January 26 1996 (cited February 5, 1996). Available from: Gopher site Gopher.uiuc.edu.

Fig 2.I
Source: Screen shot reprinted by permission of the publisher, from ***Microsoft Corporation FTP Site*** (database on-line). Microsoft Corporation (cited March 6, 1996) Redmond, WA. Accessed through Ipswitch WS_FTP [computer program] Lexington, MA: Ipswitch. Available from FTP site: FTP.Microsoft.com.

Fig. 2.J
Source: Screen shot reprinted by permission of the publisher, from ***Netscape Communications Corporation Home Page*** (on-line). Netscape Communications Corporation (cited January 28, 1996). Accessed through ***Netscape Navigator*** Netscape Communications Corporation, Mountain View, CA. (Copyright 1996 Netscape Communications Corp.). Used with permission. All Rights Reserved. This page may not be reprinted or copied without the express written permission of Netscape. Available from: Wide World Web site: http:/home.netscape.com.

Fig. 2.K
Source: Screen shot reprinted by permission of the publisher, from ***Yahoo Corporation Home Page*** (on-line). Yahoo Corporation (cited January 28, 1996), Santa Clara, CA. Text and artwork Copyright 1996 by YAHOO!, INC. All rights reserved. YAHOO! And the YAHOO! Logo are trademarks of YAHOO!, INC. Accessed through ***Netscape Navigator*** Netscape Communications Corporation Mountain View, CA (Copyright 1996 Netscape Communications Corp.). Used with permission. All Rights Reserved. This page may not be reprinted or copied without the express written permission of Netscape. Available from Wide World Web site: http://www.yahoo.com.

Fig. 2.L
Source: Screen shot reprinted by permission of the publisher, from ***North Carolina State University Home Page for the Electric Power Research Center*** (on-line). North Carolina State University (cited February 1, 1996) Raleigh, NC. Accessed through Netscape Navigator (computer program) Netscape Communications Corporation, Mountain View, CA. (Copyright 1996 Netscape Communications Corp.). Used with permission. All Rights Reserved. This page may not be reprinted

or copied without the express written permission of Netscape. Available from: Wide World Web site: http://kelley.ece.ncsu.edu/EPRC/EPRC.HTML.

Fig. 2.M
Source: Screen shot reprinted by permission of the publisher, from *North Carolina State University Home Page for the Electric Power Research Center* (on-line). North Carolina State University (cited February 1, 1996), Raleigh, NC. Accessed through Netscape Navigator (computer program) (Netscape Communications Corporation, Mountain View, CA), Copyright 1996 Netscape Communications Corp. Used with permission. All Rights Reserved. This page may not be reprinted or copied without the express written permission of Netscape. Available from: Wide World Web site: http://kelley.ece.ncsu.edu/Papers/Papers.html.

Fig. 2.N
Source: Screen shot reprinted by permission of the publisher, from *CompuServe Information Manager* (computer software). CompuServe (cited March 10, 1996), Columbus, OH. Available from CompuServe, Columbus, OH.

Fig. 2.O
Source: Reprinted by permission of the publisher, from *Dialoglink Dialog* (computer program). Knight Rider Information, Mountain View, CA. Computer screen graphic provided courtesy of Knight-Rider Information, Mountain View, CA.

Fig. 2.P
Source: Screen shot reprinted by permission of the publisher, from *ERIC Document Reproduction Services Wide World Web* (on-line). ERIC Reproduction Services (cited March 10, 1996), Springfield, VA. Accessed through *Netscape Navigator* (computer program) (Netscape Communications Corporation Mountain View, CA), Copyright 1996 Netscape Communications Corp. Used with permission. All Rights Reserved. This page may not be reprinted or copied without the express written permission of Netscape. Available from the Wide World Web site: http://www.edrs.com.

Fig. 2.Q
Source: Reprinted by permission of The McGraw-Hill Companies, Inc., from *TABE Report for Forms 7 & 8* (computer program). McGraw-Hill Inc. (cited December 12, 1996), New York, NY.

Fig. 2.R.1
Source: Screen shot reprinted by permission of the publisher, from *Grammatik 5*™ (computer program). Corel Corporation (cited April 5, 1996), Ottawa, Canada.

Fig. 2.R.2
Source: Screen shot reprinted by permission of the publisher, from *Grammatik 5*™ (computer program). Corel Corporation (cited April 5, 1996), Ottawa, Canada.

Fig. 2.R.3
Source: Screen shot reprinted by permission of the publisher, from *Grammatik 5*™ (computer program). Corel Corporation (cited April 5, 1996), Ottawa, Canada.

Fig. 2.R.4
Source: Screen shot reprinted by permission of the publisher, from *Grammatik 5*™ (computer program). Corel Corporation (cited April 5, 1996), Ottawa, Canada.

Chapter 3

Fig. 3.A
Source: Screen shot reprinted by permission of the publisher, from *Microsoft Word (Version 6.0)* (computer program). Microsoft Corporation (cited February 15, 1996), Redmond, WA.

Fig. 3.B
Source: Screen shot reprinted by permission of the publisher, from *Microsoft Word (Version 6.0)* (computer program). Microsoft Corporation (cited February 15, 1996), Redmond, WA.

Fig. 3.C
Source: Screen shot reprinted by permission of the publisher, from *Microsoft Word (Version 6.0)* (computer program). Microsoft Corporation (cited February 15, 1996), Redmond, WA.

Graphic, P.67
Source: From B. Watterson, *There's Treasure Everywhere.* (Andrew and McMeel, 1996), Kansas City, KS, 163. CALVIN AND HOBBES ©1995 Watterson. Dist. By UNIVERSAL PRESS SYNDICATE. Reprinted with permission. All rights reserved.

Fig. 3.D
Source: Screen shot reprinted by permission of the publisher, from *Grammatik 5*™ (computer program), Corel Corporation (cited February 2, 1996), Ottawa, Canada.

Fig. 3.E.1
Source: Screen shot reprinted by permission of the publisher, from *Microsoft Word (Version 6.0)* (computer program). Microsoft Corporation (cited February 4, 1996), Redmond, WA.

Fig. 3.E.2
Source: Screen shot reprinted by permission of the publisher, from *Microsoft Word (Version 6.0)* (computer program). Microsoft Corporation (cited February

4, 1996), Redmond, WA.

Chapter 4

Fig. 4.A
Source: Screen shot reprinted by permission of the publisher, from *Microsoft Word (Version 6.0)* (computer program). Microsoft Corporation (cited February 4, 1996), Redmond, WA.

Fig. 4.B.1
Source: Screen shot reprinted by permission of the publisher, from *Microsoft Word (Version 6.0)* (computer program). Microsoft Corporation (cited February 4, 1996), Redmond, WA.

Fig. 4.B.2
Source: Screen shot reprinted by permission of the publisher, from *Microsoft Word (Version 6.0)* (computer program). Microsoft Corporation (cited February 4, 1996), Redmond, WA.

Fig. 4.B.3
Source: Screen shot reprinted by permission of the publisher, from *Microsoft Word (Version 6.0)* (computer program). Microsoft Corporation (cited February 4, 1996), Redmond, WA.

Fig. 4.C.1
Source: Screen shot reprinted by permission of the publisher, from *Microsoft Word (Version 6.0)* (computer program). Microsoft Corporation (cited February 15, 1996), Redmond, WA.

Fig. 4.C.2
Source: Screen shot reprinted by permission of the publisher, from *Microsoft Word (Version 6.0)* (computer program). Microsoft Corporation (cited February 15, 1996), Redmond, WA.

Fig. 4.D.2
Source: Screen shot reprinted by permission of the publisher, from *Microsoft Word (Version 6.0)* (computer program). Microsoft Corporation (cited February 15, 1996), Redmond, WA.

Fig. 4.E.1
Source: Screen shot reprinted by permission of the publisher, from *Microsoft Word (Version 6.0)* (computer program). Microsoft Corporation (cited February 15, 1996), Redmond, WA.

Fig. 4.E.2
Source: Screen shot reprinted by permission of the publisher, from *Aldus PageMaker 5.0* (computer program). Adobe Systems, Incorporated (cited February 17, 1996), Seattle, WA.

Fig. 4.F.1
Source: Screen shot reprinted by permission of the publisher, from *Microsoft Word (Version 6.0)* (computer program). Microsoft Corporation (cited February 15, 1996), Redmond, WA.

Fig. 4.F.2
Source: Screen shot reprinted by permission of the publisher, from *Microsoft Word (Version 6.0)* (computer program). Microsoft Corporation (cited February 15, 1996), Redmond, WA.

Fig. 4.G
Source: Screen shot reprinted by permission of the publisher, from *Grammatik 5*™ (computer program). Corel Corporation (cited February 15, 1996), Ottowa, Canada.

Fig. 4.H.1
Source: Screen shot reprinted by permission of the publisher, from *Microsoft Word (Version 6.0)* (computer program). Microsoft Corporation (cited February 15, 1996), Redmond, WA.

Fig. 4.I
Source: Screen shot reprinted by permission of the publisher, from *Microsoft Word (Version 6.0)* (computer program). Microsoft Corporation (cited February 17, 1996), Redmond, WA.

Fig. 4.J
Source: Screen shot reprinted by permission of the publisher, from *CorelDRAW*™ (computer program). Corel Corporation (cited February 17, 1996), Ottawa, Canada.

Fig. 4.K.1
Source: Screen shot reprinted by permission of the publisher, from *Aldus PageMaker 5.0* (computer program). Adobe Systems, Incorporated (cited February 17, 1996), Seattle, WA.

Fig. 4.K.2
Source: Screen shot reprinted by permission of the publisher, from *Aldus PageMaker 5.0* (computer program). Adobe Systems, Incorporated (cited February 17, 1996), Seattle, WA.

Fig. 4.M
Source: Screen shot reprinted by permission of the publisher, from *Microsoft Word (Version 6.0)* (computer program). Microsoft Corporation (cited February 17, 1996), Redmond, WA.

Fig. 4.N.1
Source: Screen shot reprinted by permission of the publisher, from *Microsoft Word (Version 6.0)* (computer program). Microsoft Corporation (cited February 17, 1996), Redmond, WA.

Fig. 4.N.2
Source: Screen shot reprinted by permission of the publisher, from *Microsoft Word (Version 6.0)* (computer program). Microsoft Corporation (cited February 17, 1996), Redmond, WA.

Fig. 4.N.3
Source: Screen shot reprinted by permission of the publisher, from *Microsoft Word (Version 6.0)* (computer program). Microsoft Corporation (cited February 17, 1996), Redmond, WA.

Fig. 4.N.4
Source: Screen shot reprinted by permission of the publisher, from *Microsoft Word (Version 6.0)* (computer program). Microsoft Corporation (cited February 17, 1996), Redmond, WA.

Fig. 4.N.5
Source: Screen shot reprinted by permission of the publisher, from *Microsoft Word (Version 6.0)* (computer program). Microsoft Corporation (cited February 17, 1996), Redmond, WA.

Fig. 4.N.6
Source: Screen shot reprinted by permission of the publisher, from *Microsoft Word (Version 6.0)* (computer program). Microsoft Corporation (cited February 17, 1996), Redmond, WA.

Fig. 4.N.7
Source: Screen shot reprinted by permission of the publisher, from *Microsoft Word (Version 6.0)* (computer program). Microsoft Corporation (cited February 17, 1996), Redmond, WA.

Fig. 4.N.8
Source: Screen shot reprinted by permission of the publisher, from *Microsoft Word (Version 6.0)* (computer program). Microsoft Corporation (cited February 17, 1996), Redmond, WA.

Fig. 4.N.9
Source: Screen shot reprinted by permission of the publisher, from *Microsoft Word (Version 6.0)* (computer program). Microsoft Corporation (cited February 17, 1996) Redmond, WA.

Fig. 4.O.1
Source: Screen shot reprinted by permission of the publisher, from *Aldus PageMaker 5.0* (computer program). Adobe Systems, Incorporated (cited February 19, 1996), Seattle, WA.

Fig. 4.O.2
Source: Screen shot reprinted by permission of the publisher, from *Aldus PageMaker 5.0* (computer program). Adobe Systems, Incorporated (cited February 19, 1996), Seattle, WA.

Fig. 4.O.3
Source: Screen shot reprinted by permission of the publisher, from *Aldus PageMaker 5.0* (computer program). Adobe Systems, Incorporated (cited February 19, 1996), Seattle, WA.

Fig. 4.P.1
Source: Screen shot reprinted by permission of the publisher, from *CorelCAPTURE*™ (computer program). Corel Corporation (cited February 19, 1996), Ottawa, Canada.

Fig. 4.P.2
Source: Screen shot reprinted by permission of the publisher, from *CorelDRAW*™ (computer program).Corel Corporation (cited February 18, 1996) Ottawa, Canada.

Chapter 5

Fig. 5.B
Source: Reprinted with permission of the ASME Press (1996), New York, NY.

Chapter 6

Fig. 6.A
Source: Reprinted by permission of the publisher, from Vanston, L. K., Kennedy, W. J. and El-Bardy-Nance, S. *A Facsimile of the Future: Forecasts of Fax Markets and Technologies.* Technology Futures, Inc., (1991)Austin, TX: xi.

Graphic, p. 143
Source: Reprinted by permission of author, Copyright © 1995 by Judith Martin.

Graphic, p. 154
Source: Reprinted by permission of the artist, The 5th Wave by Rich Tennant, Rockport, MA., e-mail: the5wave@tiac.net.

Fig. 6.C
Source: Screen shot reprinted by permission of the publisher, from *CompuServe Information Manager* (computer software). CompuServe (cited March 25, 1996), Columbus, OH. Available from CompuServe, Columbus, OH.

Chapter 7

Fig. 7.A
Source: Reprinted by permission of the publisher. (Vanity Fair Corporation, 1996), New York, NY.

Fig. 7.D.1
Reprinted by permission of the author. Kenneth A. Crawford (1996), Vernon, PA.

Fig. 7.D.2
Reprinted by permission of the author. Kenneth A. Crawford (1996), Vernon, PA.

Chapter 8

Fig. 8.A
Source: Reprinted by permission of the publisher from *Documentation — Presentation of Scientific and Technical Reports (ISO 5966).* International Organization for Standardization (1982), Geneva, Switzerland. Front cover.

Fig. 8.B
Source: Reprinted by permission of the publisher, from *Scientific and Technical Reports —Elements, Organization, and Design (ANSI/NISO Z39.18-1995).* NISO Press (1995), Bethesda, MD.Front Cover.

Fig. 8.C
Source: Reprinted by permission of the publisher, from Rudland, D. L., Scott, P. M., and Wilkowski, G. M. *The Effects of Cyclic and Dynamic Loads on Carbon Steel Pipe.* U. S. Nuclear Regulatory Commission (1996), Washington, DC. Front cover.

Fig. 8.D
Source: Reprinted by permission of the publisher, from Rudland, D. L., Scott, P. M., and Wilkowski, G. M. *The Effects of Cyclic and Dynamic Loads on Carbon Steel Pipe.* U. S. Nuclear Regulatory Commission (1996), Washington, DC. Title page.

Fig. 8.E.1
Reprinted by permission of the publisher, from Rudland, D. L., Scott, P. M., and Wilkowski, G. M. *The Effects of Cyclic and Dynamic Loads on Carbon Steel Pipe.* U. S. Nuclear Regulatory Commission(1996), Washington, DC. iii.

Fig. 8.E.2
Source: Reprinted by permission of the publisher, from Hartley, T. T., Lorenzo, C. F., and Qammar, H. K. *Chaos in a Fractional Order Chua System.* National Aeronautics and Space Administration (1996), Linthicim Heights, MD. Report document page.

Fig. 8.F.1
Reprinted by permission of the publisher, from Rudland, D. L., Scott, P. M., and Wilkowski, G. M. *The Effects of Cyclic and Dynamic Loads on Carbon Steel Pipe.* U.S. Nuclear Regulatory Commission (1996), Washington, DC. vii.

Fig. 8.F.2
Reprinted by permission of the publisher, from Rudland, D. L., Scott, P. M., and Wilkowski, G. M. *The Effects of Cyclic and Dynamic Loads on Carbon Steel Pipe.* U.S. Nuclear Regulatory Commission (1996), Washington, DC. xi.

Fig. 8.G.1
Source: Reprinted by permission of the publisher, from Saltzman, E. J., and Ayers, T. G., *Selected Examples of NACA/NASA Supersonic Flight Research.* National Aeronautics and Space Administration (1995,) Edwards, CA. 5.

Fig. 8.G.2
Source: Reprinted by permission of the publisher, from Saltzman, E. J., and Ayers, T. G., *Selected Examples of NACA/NASA Supersonic Flight Research.* National Aeronautics and Space Administration (1995), Edwards, CA. 6.

Fig. 8.G.3
Source: Reprinted by permission of the publisher, from Saltzman, E. J., and Ayers, T. G., *Selected Examples of NACA/NASA Supersonic Flight Research.* National Aeronautics and Space Administration (1995, Edwards, CA. 7.

Fig. 8.G.4
Source: Reprinted by permission of the publisher, from Saltzman, E. J., and Ayers, T. G., *Selected Examples of NACA/NASA Supersonic Flight Research.* National Aeronautics and Space Administration, (1995) Edwards, CA. 8.

Fig. 8.G.5
Source: Reprinted by permission of the publisher, from Saltzman, E. J., and Ayers, T. G., *Selected Examples of NACA/NASA Supersonic Flight Research.* National

Aeronautics and Space Administration, (1995), Edwards, CA. 9.

Fig. 8.H.1
Source: Reprinted by permission of the publisher, from Rudland, D. L., Scott,P. M., and Wilkowski, G. M. *The Effects of Cyclic and Dynamic Loads on Carbon Steel Pipe.* U.S. Nuclear Regulatory Commission (1996), Washington, DC. 6.1.

Fig. 8.H.2
Source: Reprinted by permission of the publisher, from Rudland, D. L., Scott,P. M., and Wilkowski, G. M. *The Effects of Cyclic and Dynamic Loads on Carbon Steel Pipe.* U.S. Nuclear Regulatory Commission (1996), Washington, DC. 6.2.

Fig. 8.I
Source: Reprinted by permission of the publisher, from C. G. Wellemeyer, et al. *Final Report on Nimbus – 7 TOMS Version 7 Calibration.* National Aeronautics and Space Administration (1996), Linthicium Heights, MD. 47.

Chapter 9

Fig. 9.A.1
Source: Reprinted by permission of the publisher, from *Work Instructions Manual: Statistical Techniques.* Advanced Accessory Systems (1995b), Shelby Township, MI. 1.

Fig. 9.A.2
Source: Reprinted by permission of the publisher, from *Work Instructions Manual: Statistical Techniques.* Advanced Accessory Systems (1995b), Shelby Township, MI. 2.

Fig. 9.A.3
Source: Reprinted by permission of the publisher, from *Work Instructions Manual: Statistical Techniques.* Advanced Accessory Systems (1995b), Shelby Township, MI. 3.

Fig. 9.A.4
Source: Reprinted by permission of the publisher, from *Work Instructions Manual: Statistical Techniques.* Advanced Accessory Systems (1995b), Shelby Township, MI. 4.

Fig. 9.A.5
Source: Reprinted by permission of the publisher, from *Work Instructions Manual: Statistical Techniques.* Advanced Accessory Systems (1995b), Shelby Township, MI. 5.

Fig. 9.A.6
Source: Reprinted by permission of the publisher, from *Work Instructions Manual: Statistical Techniques.* (Shelby Township, MI: Advanced Accessory Systems, 1995b): 6.

Fig. 9.B.1
Source: Reprinted by permission of the publisher, from *Standard Operating Procedure: Preventative Maintenance.* Advanced Accessory Systems (1995a), Shelby Township, MI.1.

Fig. 9.B.2
Source: Reprinted by permission of the publisher, from *Standard Operating Procedure: Preventative Maintenance.* Advanced Accessory Systems (1995a), Shelby Township, MI.2.

Fig. 9.D
Source: Reprinted by permission of the publisher, from Documentation – Numbering of Divisions and Subdivisions in Written Documents (ISO 2145). International Organization for Standardization (1978), Geneva, Switzerland. Front cover.

Chapter 10

Fig. 10.D.1
Source: Reprinted by permission of the author, from Phillip R. Pellette, *Resume of Phillip R. Pellette.* (1995) New Orleans, LA. 1.

Fig. 10.D.2
Source: Reprinted by permission of the author, from Phillip R. Pellette, *Resume of Phillip R. Pellette.* (1995) New Orleans, LA. 2.

Fig. 10.E
Source: Reprinted by permission of the author, from Leon, J. L. *CDS Catheter Forming Machine Schedule* CDS, (1996) Townson, MD.

Chapter 11

Fig. 11.A.1
Source: Reprinted by permission of the publisher, from *Alabama River Newsprint Training Module: Headbox (PMA-05).* (Alabama River Newsprint Company, (1993b), Perdue Hill, AL. Title page.

Fig. 11.A.2
Source: Reprinted by permission of the publisher, from *Alabama River Newsprint Training Module: Headbox (PMA-05).* (Alabama River Newsprint Company, (1993b), Perdue Hill, AL. 1.

Fig. 11.A.3
Source: Reprinted by permission of the publisher, from *Alabama River Newsprint*

Training Module: Headbox (PMA-05). (Alabama River Newsprint Company, (1993b), Perdue Hill, AL. 2.

Fig. 11.A.4
Source: Reprinted by permission of the publisher, from *Alabama River Newsprint Training Module: Headbox (PMA-05)*. (Alabama River Newsprint Company, (1993b), Perdue Hill, AL. 3.

Fig. 11.A.5
Source: Reprinted by permission of the publisher, from *Alabama River Newsprint Training Module: Headbox (PMA-05)*. (Alabama River Newsprint Company, (1993b), Perdue Hill, AL. 14.

Fig. 11.A.6
Source: Reprinted by permission of the publisher, from *Alabama River Newsprint Training Module: Headbox (PMA-05)*. (Alabama River Newsprint Company, (1993b), Perdue Hill, AL. 20.

Chapter 12

Fig. 12.A
Source: Reprinted by permission of the manufacturer. International Envelope Company, (1995), Exton, PA.

Fig. 12.B
Source: Reprinted by permission of the manufacturer. International Envelope Company, (1995), Exton, PA.

Chapter 13

Fig. 13.A
Source: Screen shot reprinted by permission of the publisher, from *CompuServe Information Manager* (computer software). *Engineering Automation Forum Library Sections* (database on-line). CompuServe (cited April 22 ,1996), Columbus, OH. Available from CompuServe, Columbus, OH.

Fig. 13.B
Source: Screen shot reprinted by permission of the publisher, from *CompuServe Information Manager* (computer software). *Engineering Automation Forum Library Sections* (database on-line). CompuServe (cited April 22 ,1996), Columbus, OH. Available from CompuServe, Columbus, OH.

INDEX

Date Due
